THE HOLLOW EARTH
Revisited

THE HOLLOW EARTH
Revisited

Danny L. Weiss

The Hollow Earth—Revisited
Copyright © 2013 by Danny L. Weiss

All rights reserved. No part of this book may be reproduced, stored in a retrieval system, or transmitted, in any form or by any means, electronic, mechanical, photocopying, recording, or otherwise, without the written prior permission of the author, except in the case of brief quotations embodied in critical articles and reviews.

Edited by David Yanor

Cover Design by Jessica Moreno

Printed in the United States of America

ISBN 978-1-935914-26-6

To order additional copies please visit:

www.riversanctuarypublishing.com

or contact:

The Hollow Earth Research Society (HERS)
for quarterly newsletter subscription, reports, books, DVDs
and the latest research and developments
at

www.hollowearthresearch.org

RIVER SANCTUARY PUBLISHING
P.O Box 1561
Felton, CA 95018
www.riversanctuarypublishing.com
Dedicated to the awakening of the New Earth

∽ DEDICATION ∾

This book is dedicated to my daughter, Zephera, my wife, Irina, and my two sons, Joshua and Elusha.

Zephera endlessly unfolds the knowledge of the deeper mystical entities of life. She has withstood the radical tests of time by the treacherous karmic winds from ancients past to the present through many trials of the mind, body and soul. She carries the secret and mysterious pathway of the mystical Christ within her. When one looks directly into her eyes, one can only see the reflection of self. Her soul dances between the mysterious lights of the Ark of the Covenant and the presence of our beloved Creator.

She is the secret knowledge and pathway through the Tree of Life as a gateway and guardian to Paradise Earth.

To my beautiful wife, Irina—so loving and devoted. She continues to share with me an inner foundation of growing understanding. Our hearts, minds and souls have melded. To be fully human and in love! We hold hands, walking enthusiastically to the open door of wisdom's light.

My two sons, Joshua and Elusha, arrived with a mission in this life—to gently grasp the torch from our hands—to complete mankind's secret destiny and mystical journey of freedom.

Gerardus Mercator (March 5th, 1512 – December 2nd, 1594), the sixteenth century Flemish cartographer and geographer, created this map showing a huge continent lying in the vicinity of the North Pole. The land is an archipelago composed of several islands divided by deep rivers. A mountain sits in the center of the land. According to legends, the ancestors of Indo-Europeans lived near Mount Meru.

The map begs the question: How did that land mass appear on the chart? There had been no exploration whatsoever and thus, no public information regarding the arctic regions during the Middle Ages. Researchers believe Mercator used an ancient chart, the one that is mentioned in his letter of 1580.

This startling map shows a continent located in the center of the Arctic Ocean, pictured here, ice-free.

ACKNOWLEDGMENTS

I would like to acknowledge the assistance of many people, without whom this work would not have transmuted from my head onto this paper/digital creation.

My father revealed to me a profound secret of eternal life. His eternal memory displays his face before me as a golden mask, between the beginnings of ancient Egypt and the most high.

My mother showed me the reflection of the greatest beauty and the presence of her eternal youth; she stands before the gate of paradise.

My brothers, David and Fred, immortalized the cosmic deities of all time, the karmic struggles and battles of consciousness between kings of the world and their nations, while being raised into an unknown deity by forces beyond human comprehension.

Ritter von X had been generous in his contributions. I reunited with him in my most recent mystical past-life memory that empowered me with many secrets and knowledge of the Holy Grail.

I am grateful to my immediate family members and friends who shared the journey with me in the midst of chaos, beauty, mystery, aliens, exploration, research, adventure and travel. They are heir to rich, indelible memories.

Wanda Hopkins was my first confidante who helped me organize and cull through my notes, research and memos. She stayed the course with me through the first draft.

Thanks to Mark Schwartz, the producer and writer, Wendy Armagnac, and Gary Armagnac, the Shakespearian actor who narrated the DVD, Journey to the Hollow Earth.

And finally, a special thanks to the editors, Arlene Prunkl and David Yanor, who edited and finalized the last touches; I could not have finished without them. Thank you.

TABLE OF CONTENTS

Chapter One	CURIOSITY PIQUED	1
Chapter Two	EARLY VISION	4
Chapter Three	PULLING FREE	18
Chapter Four	THE UNKNOWN SPEAKER	36
Chapter Five	A NEW PATH	42
Chapter Six	THE EARTHQUAKE	53
Chapter Seven	CONVERSATIONS WITH RITTER VON X	70
Chapter Eight	CONVERSATIONS... Day 2	89
Chapter Nine	WHAT DOES IT ALL MEAN?	140
Chapter Ten	ANCIENT MEMORIES	161
Chapter Eleven	FROM OUT OF THE DEPTHS	174
Chapter Twelve	THE DREAM	190
Chapter Thirteen	LILY	204
Chapter Fourteen	GOD IN OUR MIDST	217
Chapter Fifteen	THE QUEST FOR INITIATION	226
Chapter Sixteen	THE SECOND DEATH	249
Chapter Seventeen	BREAKING THE CHAINS	255

Chapter Eighteen	FIRE AND ICE	260
Chapter Nineteen	FUTURE LIFETIME	268
Chapter Twenty	ULTIMA THULE	281
Epilogue		293
Illustration Credits		318

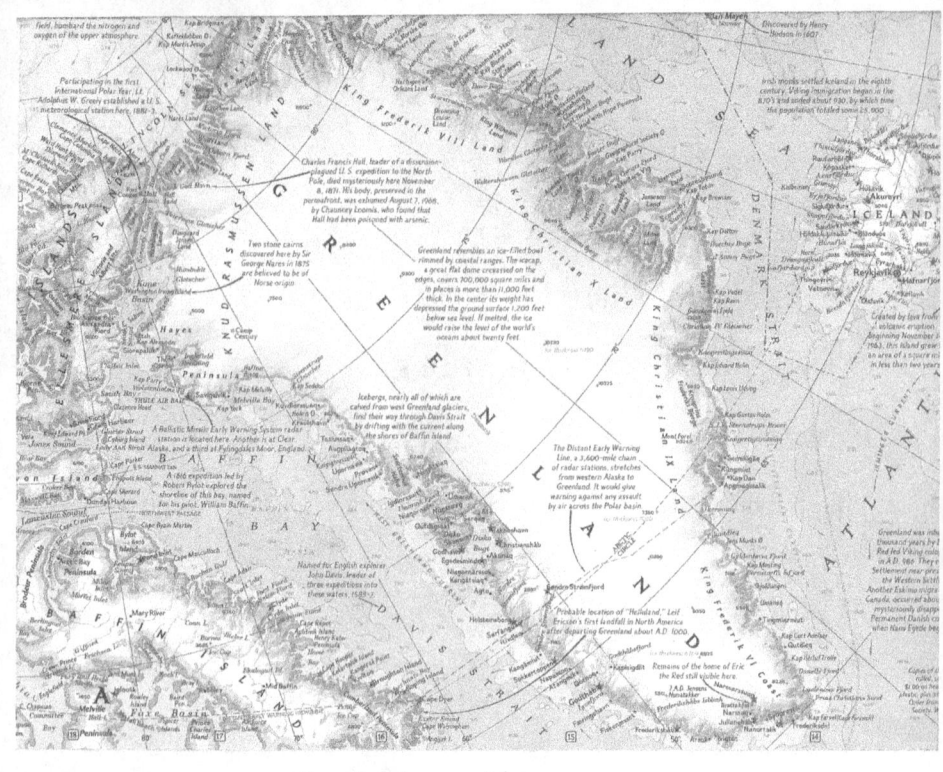

Greenland map with ice.

Greenland without ice.
Greek mythology tells us that to reach the land of the gods, one must travel far to the north, past the land shaped like a musical harp or lyre.

Chapter One

❧ Curiosity Piqued ☙

*One doesn't discover new lands
without consenting to lose sight of the shore
for a very long time.*

—Andre Gide (1869 - 1951)

Dan thought it was all so surreal—the research, the books he had discovered, his own unpredictable journey thus far. A major coup had been his interview with a secretive German who was convinced there were entrances to the interior of the earth. This older man, with knowledge of Nazi interest, claimed to have talked to one such denizen, a being by the name of Thal. Dan recalled a small slice of their conversation.

Dan asked Ritter von X another question about his experience with the hollow earth and its inhabitants.

X: This is a strange thing. Of Christ, we believe that the Bible is a combination of ancient, archaic record, written in the language of that time, as it was understood in that time. And we believe that this man, Christ, was a Prince that came from the inner world to the outer world because of the great knowledge He had that astounded humankind. You know, it is written how He confounded even the most learned and wise at that time. These men, these priests in the temples, were astounded when they heard this young boy speak, and what He knew. You know this, according to the Bible. It says Jesus died, then resurrected—he was the Son of God.

He also contained in his body a magnetic power to refurbish himself. These types of beings cannot be killed. And this is the resurrection. He rose, and this great rock from the tomb, weighing many tons, was rolled away by what some soldiers in their presence said were angels. They were beings—glowing beings. They had come from the inner realms to rescue Jesus. It was a magnanimous gesture by one of the great beings of the inner realms who came forward to try to help the world.

And since that time there has not been one, not one bit of help. Not in the sense this teacher came forward. We believe that He was a member of the inner realms who came to help—a type of prince—and He spoke of such strange things—"in my Father's house are many mansions." He had a vast working knowledge of astronomy, of religion of its time, of politics. He had a grasp uncommon to an ordinary human and the ability to lay hands upon people and literally raise them from the dead, because they possess magnetic energy. They can do anything, and we don't know how it works. We can't—we can't comprehend. But it is there.

Thal wears a dark blue mesh type of garment. It is tight fitting and he is well over six feet. A beautiful glorified body, like no human can have this kind of body. You cannot touch him. He warns you of this great danger because there's a vibration of light all around him, and the spacecraft that he comes in, it isn't anything that a man can comprehend.

You think that the aircraft you arrived in today was fantastic and beautiful. But it is a metal, riveted, put together with our technology. But the Flugelrads [UFOs] are something unbelievable. The controls inside, I have never seen anything like them. There are no levers, no wheels, nothing of that nature. Everything is a rose-colored panel of light. And to get things to work is merely to pass the hand in certain directions, to say certain things. There are no doors on these crafts—not like open this way [he gestures], or a hatchway like on a U-boat, or anything. When they land, there is nothing you can see until this pulsating light stops, a voice speaks forward to you to stay clear until the lights on the outside have ceased. And then all of a sudden there's a door—not a door, it's an opening. It defies any technological explanation. These crafts are not large. Some of them might be but—

Dan: You have seen or experienced one?

X: Yes, I have seen and have been aboard one, in the countryside. I won't say where. But I thought that night we went there, I thought, hah, this will be something very funny, something that I will ridicule this person about. It was very early in the morning, heavy dew on the grass, on the meadow, with lots of stars. It was a warm summer in August and he said that we have not long to wait. And then in the distance I saw this little tiny orange light sitting up there. And it got closer and closer, and as it came down, it became huge but it was not a bright light of any kind. It was a very soft orange glow, and it—frankly, to place it mildly—it scared the hell out of me.

I was ready to run, and I'm not a man who frightens easily. But that frightened me, scared me, terribly. And the dog that I had with me ran away, quickly. He was terrified. We found him two days later in the woods, scared to death of something. Anyway, the craft landed and spoke perfect German dialect, a very high type. This magnificent being told us to come aboard, [so] we entered, and it was not a bright burning light in there, but a soft light. It didn't hurt the eyes. I've never seen anything quite like it. But we were able to see everything perfectly without the harsh glare of illumination of this type. The first thing that I am seeing is no instrumentation, no controls. Nothing I could even recognize.

It was so fantastic in nature that I can say only one thing to myself: "I'm dreaming. This is not real. I am dreaming and this is not real."

And Thal turned and he said, "You are fully awake and I have a greeting from Dr. Franz Philip."

And I said, "Franz is dead."

"No, he isn't. He sends you greetings."

And he said something then—I knew immediately it was really from Franz because it was something very personal and private to me. I told no one about that for a very long time and I don't tell anybody, anyone, any people, that I am immediately associated with in everyday life."

And, to think—these revelations were just the tip of the hollow iceberg.

Chapter Two

EARLY VISIONS

My God, my God, why hast thou forsaken me?
Why art thou so far from helping me
and from the words of my roaring?
—Psalms 22:1

My God—when is this all going to end?

FLASHING THROUGH THE redwoods, the early morning sun blinded me momentarily, creating patterned shadows across my outdoor deck and causing me to furrow my brow, squint, and shade my eyes. A physical lump of anguish rose in my chest. I waved my hands in circular motions and paced back and forth, the sun alternately blinding and shading my face. The physical pain grew. Through the French doors, I watched as my neighbor and occasional assistant, Diane, stared at me in obvious disgust. I stopped flailing my arms, turned sharply, and propelled myself back up the deck stairs and into the house. I poured another cup of coffee.

When will there be an end to this? I asked myself, shaking my head. That's all I want to know. What the hell are we doing with this, anyway?

Diane moved to the doorway and stared blankly into the sun. This time she would leave. This time she would not come back. She would gather her books, papers, notes and tapes and *my* books, papers, notes and tapes —everything. Even though she knew they were not really hers to take.

I paced with my coffee, studying Diane cautiously as she stared into the redwoods, knowing what she was thinking. It did not matter. The sun's rays streamed over her face, and her stare, mirrored by my own, was a tangible light link between us, like a laser beam growing brighter, brighter, brighter. My ears were screaming now, and I felt almost paralyzed. I had a sudden flashback to a time when, as a child, I used to design charts by the hour—maps of a world unknown to me, but familiar somehow, compelling me to continue. The hidden and ancient memory had returned—the memory that drove me to exchange childhood games for charts tangled like spiders' webs, curving upward to a center, overlapping, defining the perfect rim. Hour after hour I would make plans for a journey I was never able to define; my mother finally gave up her questioning.

I stood now, frozen in a shaft of light extending out to Diane, my entire being filled with high-pitched tones that demanded, "Tell it. Tell it all. Tell it now. Now is just the beginning."

Diane, stoic and mechanical, turned, walked back into the house, and gathered her things, rather, my things. I followed after her inside, watching as she leafed through my papers and tapes. Boldly she selected exactly what she wanted. She turned toward me.

"Ritter von X is right, you know. I've been coming to that conclusion for a long time, and finally it is clear, very clear. We must each do our own work now, find our own paths. There's not much time, and I will have to move fast—on my own. You understand, I hope?"

Distracted, I ran my fingers through my thinning hair and said nothing. I was aware of her movements, but unconcerned. As she left, there was a noticeable shift in the colors, shapes, movement, and sounds in the room. The sun still streamed in through the front windows, a luminescent green glow filled the room, and the screaming tones in my ears transformed into sounds of violins on the wind—Wagner's "Ride of the Valkyries." I pictured the shielded maidens with golden hair and snowy arms carrying the slain off the battlefields to Valhalla.

I remembered an incident from my childhood. I was six years old. I was lying in my bed, focusing on the quiet, pulling the blankets up to my neck. My attention slipped outside, where the wind was rocking the giant redwoods against one another. The creaking sound made me think of gigantic robots moving past a distant waterfall. I felt the quietness—the stillness. I was the quietness and the stillness.

"I am the stillness." Suddenly I was frightened. Quickly, I opened my eyes and immediately felt safe and reassured by my familiar surroundings. I returned my attention to the silence, and the stillness poured through my body. I thought of nothing. I began slipping into what seemed to be a deep sleep, entering a vast spacious opening, feeling something huge and awesome. I was immersed within an endless being, breathing in its supreme enormity.

I had mentioned the surreal experience to Wanda.

I was in a large auditorium. I felt myself sliding into a small room where I became me—the child. I had touched what seemed to be a larger part of my own self. I felt at one and yet, incomplete in some mystical way.

"You mean you actually think you were there?" My neighbor's voice on the other end of the phone was skeptical. She paused and then challenged, tentatively, "I mean, weren't you frightened?"

"Wanda! You don't believe me."

"I didn't say that, damn it! I just wondered if you really thought you were somewhere else—there—wherever you were, I mean. I don't know—the center of the Earth or something? I don't know what you thought—that's what I'm trying to find out!" Wanda's voice sputtered to a stop. She sounded agitated and frustrated.

I continued trying to explain. I was visualizing and revealing ideas to myself as I spoke.

I waited for some intuitive impression or direction or confirmation from my higher self." "Then an intense vibration quivered throughout my body that slowly increased to a barely audible, distant sound. The sound grew louder and then very intense. My higher self seemed boundless. For a few seconds I lost touch with my childlike self, and I became frightened. I couldn't move or speak.

"I guess not." Wanda sounded incredulous. "Did you talk to anybody? Was there anyone else there?"

"Not really, not then—that came years later. No other person was there in the flesh, then, but yes, there were several impressions of some things, some ones, but they didn't speak to me—not that first time. Anyway, suddenly I crashed back into the silence of my bedroom and opened my eyes. I was in my bed. I sat up, feeling as though I was being swallowed by an enormous being that was filled with an eternal stillness of a profound peace and an exalted sense of freedom. I was slipping into another universe that knew no boundaries."

"And this was when you were six?"

"Yeah, this stuff has been going on forever. So anyway, I sprang from my bed and ran into the living room where my mother and father were watching television. Frantically, I told them that there was something gigantic in my room. I didn't mention, or even try to describe, the ineffable sensation." I spoke the last words with heavy breaths and emphasis. "My mother was very compassionate, trying to explain that it was just a dream. Eventually, I felt comforted. She walked me back to my bedroom and reminded me to say the Lord's Prayer. So I began to recite the prayer, but fell asleep before I had finished it."

"Good grief."

"I slept comfortably and the next morning awoke refreshed with a feeling of a new kind of strength and vitality."

"Well—" Wanda paused. "I've always known I lived next door to an interesting kind of guy."

"Yeah, well. Then, at the age of thirteen, I stepped out of my body. It was a curiously normal and somewhat familiar feeling. When I'd completely risen and separated from my body, I moved—"

"Listen," Wanda cut me off. "I got to get going, you know. Sorry. Talk to you later. General Hospital is on in a few minutes. And I've got to put up my hair before Max gets home. Talk to you later."

Part of me heard Wanda hang up the phone, but the story continued telling itself in my mind.

I had moved swiftly toward the oval mirror that hung upon my bedroom wall. My physical body was there on my bed. Dumbfounded, I stared into the mirror. Instead of my own image, I found myself facing a bearded man who was laughing loudly. For a moment, he held back his throaty laughter and spoke to me.

"There is no William Shakespeare!"

Again, he returned to his laughter, and then his image disappeared from the mirror. I slammed back into my body, opened my eyes, and sat up in the silence of my bedroom.

Who is William Shakespeare?

I wasn't introduced to him until later in school. Years after that, I learned of the theory that Sir Francis Bacon, who, some say, was the embodiment of Saint Germaine, may have been the real author of Shakespeare's works (information not trumpeted at school, needless to say). But the man in the mirror and his strange words only left me shrouded in a dark mystery. These were not subjects in the normal junior high curriculum; school became meaningless to me. Perhaps, had any teacher spoken of the soul, our purpose on this Earth, the meaning of life, or any mystical phenomena, I may have been inspired to discover the noblest ideas early in life. But school was merely too long, too drawn out, too superficial, offering nothing of what I really needed on an emotional or psychic level.

An inner part of me was being awakened, yes, but certainly not in school. On the contrary, school suppressed my inner self. True learning had to be done on my own. My own inner teachers had to be summoned in order to reveal the mysteries I was experiencing. My internal dialogues became as real as any living experiences and far more educational than any junior high classroom.

One night soon after my out-of-body experience, I was awakened by a brilliant white circular light in my doorway. The light entered my bedroom and hovered at the foot of my bed. Immediately, the light moved through my body and up through my head, covering my face completely. It felt like a strong, whirling wind on my face. The light then receded in the same way it had approached me, leaving my body behind. Finally, the light and I—my soul—hovered near the ceiling for a few moments, where I felt paralyzed between two worlds.

Today, reflecting back, I think then I was being taught some mysterious lessons. I often imagine a gathering of all the people in my research. What if they were to meet right here in my dining room and have a conversation? What would they say? I wonder about that sometimes—a gathering of great cosmic voyagers. I would serve tea, perhaps, and just eavesdrop on the conversation.

President Pierce, of course, would be the moderator. I do not know why, but I have always felt an affinity with him. He has been like a friend to me over the years. Sir Francis Bacon, the bearded man in my mirror, would probably say a few words. Dr. Edmond Halley, after whom Halley's Comet was named, should be there, too. Just imagine it.

Dr. M.L. Sherman and Professor Lyon, writers (in 1871) of The Hollow Globe, could be guest lecturers. William Reed, Marshall B. Gardner, Ferdinand Ossendowski, Captain Cook, Dr. Raymond Bernard, and even Admiral Richard E. Byrd along with his lieutenant commander, David E. Bunger, might even tell me what really happened to the United States Navy's mission, "Operation High Jump" at the South Pole in February of 1947. Professor Lyon would probably talk too much. What a magnificent gathering that would be.

Seventeenth-century astronomer, Edmond Halley, of comet fame, holds a diagram of his hollow earth concentric ring theory.

"Then it is by no means strange that philosophers of the present day have not developed a very clear conception of all the causes that have operated in producing the formation of the globe upon which we dwell; that they should be somewhat mystified in relation to its physical arrangements; that they should also adopt theories that are so very unnatural and antagonistic to well-defined principles pervading the universal realms; and that eminent scientific men have acknowledged the existence of—"

"My dear Professor Lyon, please!" interrupted Pierce. "We must maintain some kind of systematic order here. As your duly elected president, and in concern for our young guest, so as to allow him the greatest benefit of our collective knowledge, we simply must not indulge in expounding endlessly with these purely self-interested topics. Please, let us stick to the subject at hand. You and Dr. Sherman's lofty theories regarding a mechanical hollow globe, which were most eloquently put forth as early as 1871, can be discussed at a later gathering. Tonight's topic concerns itself with William Shakespeare, not the Hollow Earth." Clearing his throat and adjusting his tie, President Pierce continued, "Now, let me see, who will be next? The floor recognizes Sir Francis Bacon, the Imperator of the Rosicrucian Order."

"Of course, and thank you gentlemen, and our most honored young guest this evening. Thank you for attending. Now, anyone who has referred to the King James Version of the Holy Bible has no doubt performed this simple test. I challenge you, young man, to follow these instructions precisely. Please turn to Psalms 46 and count forty-six words down from the top of the psalm, and circle the forty-sixth word. Then count forty-six words from the bottom of the psalm, but do not count 'Selah' as the first word at the bottom of the psalm. You will most certainly arrive at something very interesting. You may ask yourself, why the number forty-six? Aha! Yes, well, after you have put these two circled words together, you will have a logical clue to the number forty-six."

Concluding his instruction with very little direction as to exactly what we might discover, Sir Francis Bacon promptly disappeared from my imaginary gathering, while I continued, again lost in the memory of the brilliant white light.

Back in the bedroom of my youth, the brilliant white circular light that had captured my soul was receding in the same way it had approached me, leaving my body behind. It hovered with my out-of-body self near the ceiling for a few moments, until we were vacuumed into an immense, unlimited sense of space.

At first, I tried desperately to resist its mighty power by hitting the wall next to me. My intention was to awaken and alert my parents, who were sleeping in the next room. Yet, I found it impossible to hit the wall—my hand passed directly through it. Suddenly my resistance was stilled. A low-pitched voice spoke in a rhythmic tone, resembling the sound of rolling thunder. The familiarity of this sound caused me to completely relax my struggle with the awesome, brilliant light. After this celestial communication, the light descended into my body again. The light pulled away and became smaller, then disappeared through the doorway as though it had moved through its own timeless eternity.

For several years I did not speak to anyone of my mystical experiences; I felt that it was somehow *verboten*. In fact, I had totally dismissed them from my mind and had immersed myself in my childhood activities. But the teachings remained in the recesses of my memory. It was not until a later time in my life that a glimpse of wisdom's light began to surface in my everyday thinking. But I failed when I tried to describe the origin of these cosmic encounters using the dull tools of my intellect alone.

During my teenage years, I lived a hidden life of duality. One part was blinded by these cosmic blasts of light and their overwhelming sense of freedom. The other wore the cloak of its time, dressed by the subtle influences of my earthly identity. I was very shy. But I was always grateful that my parents gave me the freedom to think and decide things for myself. It was important to them that I make my own decisions and set my own limitations, goals, and boundaries.

I grew up thinking I was meant to be a scientist. I could not find enough books on scientific subjects and I hoarded any that I did find. And while I had two older brothers, I usually played alone. I was constantly designing and engineering multistory missiles and had an inexplicable desire to climb to the stars. I was influenced by the era in which astronauts first touched the fringes of outer space. I found myself preparing and mixing volatile and dangerous fuels for combustion and confined myself to a home laboratory for hours on end. This is where I dared to put my dreams into reality. In retrospect, it was fortunate that no one was killed in my lab. Even after a few major accidents that nearly ended in disaster, my parents still trusted me when I was alone in my home lab. Hiding my thoughts and creativity, I remained in a silent, nebulous world, puzzled, yet strangely satisfied with the occasional midnight visitations. I remained content within my lively world of the unknown.

One night, my father and I were sleeping outside our house, which was not uncommon for us during the warm summer months. It was very late and we had still not gone to sleep. I was awestruck and intrigued by the vast night sky and its perfect, yet chaotic order.

"Could other people live out there, Dad?" I asked.

"Would it make much sense if we were the only human beings in all existence?" he asked. He wanted to know what I had just been thinking.

"There's got to be some other kind of people out there, huh, Dad?"

I began to doze and close my eyes while occasionally glancing back at the night sky until I could no longer focus on the mysterious unanswered questions of the universe.

"Look there, son." My father drew my attention to an object in the sky that was heading south.

My eyes flew open, and I saw the bright object he was pointing at. "That's a satellite!" he declared with solemn conviction, "because it's moving on a horizontal line. See how it seems to tumble and fall along a jagged trajectory?"

I was enthralled by the silence and the clear night sky. Subtle winds surrounded us like waterfalls rustling through the tall redwood trees. It all remained a mystery to me, and I soon drifted into a deep, peaceful sleep.

Some time lapsed. Suddenly, I was awakened by beautiful orchestral music. I was filled with this wondrous celestial composition; it was as though the whole universe was echoing a song of profound peace and harmony. I focused my attention to the north and beheld a saucer-shaped disk, soon realizing that the music was emanating from this strange object. It was metallic in appearance with bands of reddish-orange color across its surface. As it moved eerily across the scintillating night sky from north to south, it sounded like a symphony an orchestra of musicians playing an infinite array of instruments in harmonious octaves. The saucer was tilted in a position so that I could see the top view as it continued to pour forth its heavenly sounds. Immediately after the saucer cleared my view, another one emerged from directly behind it, suspended at almost the same angle as the previous one. This one was slightly tilted so that the underside was exposed with only a slight view of its crown top. It was also metallic in color, and it spun on an invisible axis as it moved across the starry heavens toward the south. The same enchanting music issued from its path, and a great sense of peace enveloped me. The orchestration from this craft was different from the first. It was like the drawing of a violin's bow across and back, producing varying pitches but heard as one continual stroke. However, it was the same harmonious arrangement of sounds. The symphonic vehicle passed before me, then out of sight and earshot, leaving me awestruck while my father slept undisturbed through the night.

In the wake of my reverie of saucers and symphonies, a raspy voice in a guttural, thick German accent pierced my consciousness. It was a voice from the future, where, it seemed, I had just traveled. In this vignette, I was at home, listening to a cassette tape from Ritter von X:

"...And there is a great storm brewing in the world. It is between the forces of darkness and the forces of the light. Terrible as it will be when it comes, you can rejoice in the fact that the world will be purified and cleansed of all evil and soulless ones, and all the powers of darkness that reside in them. You will see arising from the Great War that is coming, a new world, and a bright world. Then the true nature of man will be at last restored to the divine nature that he once possessed."

Next, a distinct upper-crust English accent rang in my head as I read from Dr. William F. Lyon's 1871 book, *The Hollow Globe*:

"Man is a microcosm, containing all elementary constituents found upon our globe. With vision, man can behold in man or in one drop of water, all that there is to behold. Universal vision will reduce the vast ocean to a single drop, or the globe we inhabit to the dimensions of a grain of sand, and peer out upon the remotest orb that has come within the range of the most powerful satellite, and bring it within hailing distance, and hold easy converse with its inhabitants. Our very thought particles are

distinctly visible to advanced spirit intelligences, as the flowers that adorn our gardens or jewels that adorn our necks and fingers. How shortsighted is humankind! How narrow the limits of human vision! We are one."

Suddenly, I was slammed back in time to the beautiful symphonic music I had been hearing. My eyes opened suddenly. Was it just a dream, or were the voices I had just heard someone or something yet to materialize? It was frightening—yet attractive, intriguing, and as awesome as the universe itself.

Once again, I was enveloped by the silence. My head felt heavy. I could no longer hold the craft in my eye as a fixed position among the stars. I began to slip into a deep and relaxed sleep.

In the days that followed, I remained quiet about my experiences. But later, feeling the urge to share, I mentioned that I'd had a dream about two flying saucers passing through the night sky while making a symphony of music. Its strange and wonderful music remained in my mind, and I found myself compelled to look up the word *universe* in the dictionary. When I found its meaning, my feelings were a strange combination of surprise and confirmation of what I already seemed to know at a subconscious level. The prefix *unus* means "one" and *versus* means "song." What song, I wondered, would all living beings in the universe sing as one?

I felt that somehow, America's quest in space might yield information to answer that question. I was an ardent follower of the first American astronauts and Russian cosmonauts. On occasion, I would stay home from school and watch their orbital TV broadcasts and reports from the NASA Space Center. I knew our knowledge was very limited concerning space travel and other solar systems and their planets. It was my own belief that other planetary beings existed and were well advanced compared with Earth's creatures. However, as a child, that was mostly just an idea residing in my imagination and those of my friends. But I devised an experiment.

One evening, I went outdoors with my mother's old pop-out Kodak camera. Beneath the night sky, I set up the camera to take some pictures of the full moon. The setup was simple and primitive. I shot a roll of film and waited anxiously for its return, upon which I was impressed at how so many pictures came out in strange and unusual shapes. I shared the photos that I thought were most interesting with my classmates at school. I presented the most interesting one first.

One student declared confidently, "Hey, this is a flying saucer!" His voice commanded the attention of several nearby onlookers, who excitedly echoed a similar belief—the photo was real.

The trickery gave me no real satisfaction. I gave in and explained that while I was taking night pictures of the moon I had jiggled the camera slightly. When the film was developed, the moon appeared to be shaped exactly as flying saucers were reported to be. Needless to say, those who had wanted to believe it to be real were disappointed. My young friends wanted to believe in its authenticity as much as I did, but it simply was not the truth. I lost credibility among my friends for a while. Yet, they would have accepted the lie, and would have believed what I had presented as evidence of the truth.

I learned a lesson that day: I realized that to accept something by believing in it immediately, just because you want to or because it serves your purposes, limits the truth. Since that incident, I have always been extremely cautious when defining what I or others consider to be the truth.

But that did not stop me from acting out my boyish pranks. A few months later while I was visiting a friend, we decided to make several phone calls to people at random, asking them if they believed in UFOs. I also asked, "And if you do believe in UFOs, would you be interested in contributing to a research organization?"

My results were positive nearly all the time. Some related stories of their own sightings—of what they thought might have been strange UFO phenomena. The consensus was that people believed in UFOs and simply wanted to know more of the facts and the truth.

During my last year of high school, I had a good friend, named Gary, who shared my deep interest in UFOs and space travel. In fact, we frightened our parents terribly with our conversations. Indeed, some of Gary's ideas were so outlandish that few people could comprehend them. We took these reactions as complimentary. Gary and I were very serious, and we began to design and make plans for a saucer-type spacecraft. We spent nearly all our time developing ideas for our intended saucer.

We figured our hull should be approximately twenty meters in diameter. When we came up with the idea of how to construct the hull, we considered ourselves clever. By digging a massive hole according to shape and design for the bottom of our spacecraft, we could create a mold. Then we would dig and construct another mold for the top of the craft.

After the two halves had been constructed, they would be placed together. We figured there had to be some type of metal that could be heated, melted, and poured into our mold design and that could withstand the stress needed for our purposes.

We decided some source of ion generation was our answer to power a gyro-system capable of producing an anti-gravity effect, to float and maneuver on the magnetic forces of the Earth. We thought to use the liquid metal, mercury, for gravitation guidance and control. We were always confident that we were on the right track. Later, as it would turn out, the same ion generation would be used for long voyages in space. But being sixteen years old at the time, our biggest obstacle was lack of money.

In all seriousness, we wanted to unite our individual imaginative intellects and combine them with our inner vision. We constantly laughed and pointed out all the superficial realities of our earthly world. We felt the world's limited knowledge was upside down in its thinking and picayune in comparison to the untapped knowledge of the universe.

In time, because of uncontrollable circumstances in our lives, our grandiose ideas faded. But our dreams remained. We felt that no one really understood us, and that somehow we were connected to a deeper understanding. We felt the world had alienated itself from its true source. We never lost sight of our confidence or our ability to unite our minds to design and build such a spacecraft.

One day, while out cruising in my 1952 Ford, we were drawing our ideas on the windshield and waving our arms around in a flurry of excitement. Our theories were becoming near realities, we thought. We were, however, unaware at that moment that we were being followed by the local sheriff. Suddenly, he flashed his lights and sounded his siren, finally getting our attention. He pulled us over. With the innocence of our excitement, we could not figure out why we had been stopped. "I've been following you for five minutes!" he said sternly. "Finally I had to pull you over. What the hell is going on? You've been waving your hands and making wild motions like crazy kids." The officer was polite but firm. "What could you possibly be talking about? Your attention was not on your driving, that's for sure!"

"Then—who was driving the car?" Gary asked.

We looked at each other and then at the officer. Silence fell upon us. Gary began to mutter something to do with a science project we were working on and how we had become carried away. The officer gave me a stern warning and said to keep both hands on the wheel. I complied and we went on our way.

One thing and then another interfered; our research projects were always interrupted. Our interests extended to other diversions. We had an accomplished rock and roll band and performed at school dances and parties.

Then, at the age of eighteen, we both needed to register for the military draft. We believed that this was the epitome of being human. I wanted to go to a college and become a pilot, but with lack of direction and having graduated from high school only six months earlier, I chose to enlist in the US Air Force. I believed at the time that I was heading in the right direction. I was classified l-A. The last thing I wanted was to be drafted into the US Army. I selected the career of Munitions and Weapons, and one year later, I graduated from Lowry Air Force Base Technical School in Denver, Colorado. I then spent nine months of on-the-job training at Luke Air Force Base in Phoenix, Arizona. I went home for thirty days, and then transferred to my next duty station at Phu Cat Air Base, Vietnam, in 1967. I saw Gary only once after that, for just a few brief moments.

Chapter Three

~ PULLING FREE ~

In peace, sons bury fathers, but war violates the order of nature, and fathers bury sons.

—Herodotus, 484–430 BC

DEMONSTRATIONS AGAINST THE war in Vietnam did not concern me or have any significance in my life at that time. I was very much in love. My dual focus was my young wife and my new Air Force career. In fact, little meant more to me than Patsy, my gorgeous, blue-eyed, blonde, sixteen-year-old Norwegian wife. We married four days before my departure to Vietnam. With little to hold us back, we had a beautiful wedding with all the traditional accoutrements. My father sang the Hawaiian wedding song in his magnificent tenor. A joy, mixed with a premonition of sadness, flooded over me as we departed for our honeymoon destination, ecstatic and in love. I had no idea at the time that Hawaii represented a major crossroad in my life. At my departure, the sadness I felt at leaving my new bride was unleavened by the danger and uncertainty that lay dead ahead.

Just as quickly, I was gone. I kept my new bride's sweet caresses in a secret corner of my mind and heart, longing for the moment when I would see and hold her close once again.

Upon arriving in Vietnam, I was introduced to a new circle of acquaintances and rushed to my new duty assignment. That evening, under a heavy mortar and rocket attack, I was rushed to my assigned bunker. Confused, I simply followed the crowd ahead of me. I ran through the hallway of the barracks and down the stairs, not knowing what to say or if I should scream. I did not know where I was going—I do not even remember touching the stairs. My whole body felt numb.

Men were scooping up various personal items as they made their way to what they hoped was safe refuge.

I recall one fellow grabbing a half-finished bottle of scotch while yelling, "You're staying with me the rest of the night, sweetheart!" Another airman grabbed a portrait of his wife, while yet another turned on his TEAC tape deck to record the whole event.

My quarters were on the top floor of the barracks, somewhere in the middle. It was considered to be the worst location—the longest run to safety from the highest point. After what seemed like an eternity, we finally made it to the bunker. The bottom floor was already neatly packed into the bunker. Immediately thereafter about forty of us came running from the top floor. I was caught somewhere in the middle of the crowd. We finally reached the outside and came to a screeching halt in front of our bunker. I noticed several parachuting flares being dropped from our C-130 aircraft that lit the night sky. We could hear incoming enemy rockets soaring through the sky and ripping apart everything in their path. Finally, I saw my entrance to safety. About fifteen airmen behind me were desperately pushing another twenty airmen in front of me, two and three at a time, squeezing them through the sand-bagged bunker doorway. Since I was near the end, the men behind bulldozed the crowd ahead of me into the bunker.

As I groped my way through the small entranceway, I stumbled and fell, found my head ground halfway into the sand; this was my first taste of my new home life in Phu Cat, Vietnam.

I heard someone say to me, "Weiss, did you bring the fucking beer?"

While spitting sand from my mouth, an airman handed me a bottle of scotch to wash my mouth out. "At the next bunker raid party, Weiss, bring the damn booze."

I heard another voice with a deep southern accent say, "I hope there aren't any fucking snakes in here again." Another comment that turned into a chant was, "Which way to Sweden?" Since I was a new arrival, I was given the remainder of the scotch to drink.

While becoming acquainted with my new friends, I remarked, "Skjoll, which way to Sweden?" A comforting, hilarious burst of laughter echoed through the bunker. Within minutes, my comrades and I had finished the scotch. Before an hour went by, the rocket raid was over and the all-clear signal was sounded, telling us it was safe to return.

The morning sweep revealed some Vietcong bodies—those who were responsible for the rocket attacks. I was surprised to notice that the rockets were Soviet-made. Some airmen piled the bodies on a pallet, and then carried them with a forklift to a nearby freshly dug hole. It was a quick burial. I could see they were young girls between the ages of twelve and sixteen. Little did I imagine how, in the coming months, South Vietnam and the ground where I stood would be completely overrun and destroyed. It was not until late January of 1968 when the Vietcong launched an all-out attack on their Lunar New Year holiday called the TET Offensive that carried them to a temporary victory. It was a series of surprise attacks throughout South Vietnam.

Eight months passed and I heard from my new bride only once. My humorous and positive attitude deteriorated rapidly. However, I managed to stay away from drugs of any kind; I did not even smoke cigarettes. My favorite pastimes were drinking scotch, photography, sexual fantasies, and karate training with the Korean Army. I really grew to hate the taste of scotch. Home was a distant dream and for many a fading memory. No one spoke about it until someone had at least thirty days left of their tour of duty. For various reasons, many stayed longer—up to six months and more than their mandatory tour of one year.

As a diversion, one day my roommate and I decided to remodel our living quarters. Without care, concern or permission, we began stealing everything we needed in order to make ourselves comfortable. We declared there would be no class distinction between officers, airmen, and the enemy, as far as comfort and respect was concerned. We decreed this as a law unto ourselves. We proceeded to tear down the wall between two "cubes"—the nickname for our living quarters. We put the bunk beds at one end of the cube and made it into a unique bedroom. Every month we hung the playboy centerfold, neatly framed and tacked upon the wall where it could be viewed from a prone position on our beds. This was in addition to portraits and recent photos of our wives and girlfriends. On the other end of the cube we built a bar with four bar stools. We had a potted banana tree plant next to the couch we had stolen from the officers' transit barracks. A portable TV and refrigerator were soon added. Time passed slowly. Comical events and deep emotional sadness existed side-by-side, along with the overarching instinct to survive.

One day my supervisor granted my friend, Ken, and I two bottles of Jack Daniels and a one-week leave to Singapore. This meant only one thing in our minds—a lot of steamy sex! Our flight arrived and about two hundred servicemen were herded off the plane and delivered to designated hotels. Upon our arrival, we were briefed about the rules, and then taken up to the fifth floor where our hearts began to pound with great excitement and desire. Our emotions blurred as we realized that this would be the night that our passions could run wild.

We were introduced to about a hundred charming ladies. My friend and I went to our hotel room, where we discussed what our wives would say if we spent a week with a hooker. We consumed a considerable amount of alcohol, hoping that we would arrive at some rational yet guilt-free decision. One thing was certain: if we did not make up our minds soon, the other GIs would get all the good-looking girls. With that excuse, we staggered out of our room and entered another, where the girls were waiting.

Awkwardly, we chose our girls and hurriedly returned with them to our rooms. Kim, a Chinese girl, ran me a warm bath and gently massaged me with some kind of scented oil. Then she took me to the bedroom. She kissed me softly with a smile and a strange kind of loving kindness. Need I say more? The next day she invited me to the Buddhist temple. She was a student of Buddha's teachings. I was struck by how the teachings of Buddha seemed to allow her a great sense of non-attachment and freedom. At the time, I was deeply and quietly attracted to this quality in her.

In the meantime, I knew that my wife was securely tucked away and waiting for me at home. Unquestionably, I was still deeply in love with her. At the same time, I grew to have no real interest in the Vietnam War, but I still had to deal with military rules and suppress all my inner emotional thoughts.

I received two letters from my wife during my entire stay. Some airmen were getting divorced by mail, and I began to feel haunted by this idea. Fear crept over me like a ghostly shadow. Many men died in accidents, in combat, and by suicide, but many, like me, died within our hearts. The effects of so many kinds of destruction began to grow as a collective negative energy. I was only eighteen, and my impressionable character was being built on a negative foundation—consuming bad food, drinking foul water and too much booze, and watching people being destroyed in various ways. Like everyone else, I was trying to survive—any way I could.

Ultimately, I came to believe that the United States was trespassing. Victory was not the military's mantra and the war's purpose was never clear at any time—least of all to me. I was programmed to do a job, to destroy lives, and to trespass for reasons of enormous political aggression from an unknown origin back in America.

I was assigned a three-month temporary duty assignment at Da Nang Air Base. I had injured my shoulders and left knee—torn them to the point where they would easily dislocate. We were over-worked, and suffered from lack of sleep and lots of bugs. We were constantly dodging rocket and mortar attacks throughout all hours of the night and fighting amongst ourselves. To this day, they remain a permanent injury. After my return to Phu Cat Air Base, I had a terrible time trying to explain my injuries to some inept medics. They offered me pain pills, but I refused them and left. A few days later, my knee was swollen to the point where I could no longer walk. This time I was given an ace bandage and a second offer of pain pills. Finally, I was sent to Cam Rahn Bay Hospital, where I was granted a couple of weeks to determine my injuries. The medical decision was to not operate on my shoulders and knee; instead, some of us were sent on an excursion on a navy boat to a nearby island. We were instructed to swim in the warm tropical waters in an attempt to recover by physical therapy. In a dreamy state, I sat and observed baboons scampering about on the rock-covered hills, while thinking of the lyrics to the song, "Sitting on the dock of the bay, watching the tide roll away."

After I returned to my base with some remaining time off, my good buddy, Rocky, shaved my head as a silent protest knowing that

I had mentally crossed over into another zone. Gradually, Vietnamese housemaids began to silently invade my cube and took to cooking and eating in my room. One night, as they cheerfully offered me fresh cooked rice mixed with steamed cabbage, some kind of mysterious meat, and rice-fly mixture, one of them asked, "You same—same Buddha?"

I said yes, and told them I'd been with a wonderful Chinese girl in Singapore who was a Buddhist. As I ate, I told my new Vietnamese friends that I enjoyed learning about the teachings of Buddha. All the while, a great sadness was enveloping me. I could feel the pain we were inflicting on the Vietnamese people. They would always smile through a gray cloud of humility, with a kind of optimism that came not from my world.

Vietnamese Buddha with Chu Van (swastika to most Westerners) — symbol of enlightenment and the achievement of Nirvana

Later, I found myself huddled in a corner with a wool blanket wrapped about me. I began to feel sick inside. Tears fell, but they went unnoticed, absorbed by the blanket. I forced a smile and apologized to the group of women for all the unfathomable pain that we had brought from America. I told them that someday all the Americans would just leave and most likely all at once. One of the Vietnamese girls approached me, handing me a solid gold necklace, the pendant in the shape of a swastika. She said the swastika was an ancient symbol of the power of God and told me to wear it next to my heart. I enjoyed their company, and I had a new respect for an ancient teaching. This experience contributed to my growing understanding of a greater freedom. I later acquired a solid gold heart and chain made by the Vietnamese and sent the two items to my brother and his wife as gifts.

Soon after, I was summoned to see the first sergeant regarding my odd behavior. When I reported to his office, he stormed at me, commanding in a fierce voice how I should be a soldier and pay attention to the purpose of our mission. I fell silent and obeyed his military lecture. I knew then that the war, for whatever reason was over for me, personally.

The time came when I had only two weeks left. I had made a thirty-day countdown, numbering thirty cards and tacking them to my bedroom wall. Each day throughout the countdown, I would tear a card off the wall and set it on fire. Symbolically, I watched my last days disintegrate in the purifying flames and my past crumble into ashes. I was hoping to sever war and its causes from my life for all time. I lost track of my countdown somewhere around the fifteenth day.

A rumor spread that our departure date was two weeks early. Indeed, some of my comrades received orders to that effect. When we realized that some of the men we had come over with were departing early, several of us marched in outrage into the personnel building. We approached the first lieutenant's office and immediately noticed a large stack of orders on his desk. Our group broke into a ravenous, screaming mob, insisting on getting the orders from the officer in charge of distribution. Instead, he ordered us out of his office, saying that he would be giving us the orders in a few days. One member of our group leaped from behind, pushed his way through, and grabbed a handful of the orders. He began reading off names, and then tossed the orders into the crowd. The officer demanded that we leave. He made a feeble attempt to call the security police, but we overran his office. The orders continued to be tossed into the crowd.

Then I heard my name called, "Weiss! Your orders to Sweden are here!" he said jokingly. I grabbed them in mid-flight. The lieutenant was furious. We had humiliated the man so badly that he grabbed the remaining orders and threw them into a nearby trash can. Screaming at the top of his lungs, he threatened to have all of us arrested.

Our ringleader backed him into the wall, and with the irate crowd supporting him, yelled, "The lifer is threatening to call the air police!" The persistent crowd pushed the lieutenant toward the trash can. He fell over backward, defending his job to the end. In a rage of panic, he wrestled with several others diving toward the trash can to rescue the discarded orders. The lieutenant grabbed the rest of the orders and threw them into the crowd. Our instigator finally found his orders. We all left as fast as we had entered and stormed back to our barracks, where the booze flowed copiously as we prepared to book our flight the following morning.

We left Phu Cat and arrived in Cam Rahn Bay by a C-130 transport. We entered a long line while our orders were processed for our flight back to the U.S. New soldiers were entering for the first time while we watched and waited in line for three days, inching our way back to freedom—back to where we had come from with such innocence and optimism—just one year earlier.

The war had shattered any possibility of diplomacy between the military and me. Although it was seemingly nearing an end, for me it was just beginning. The great pressure from anti-war activists almost thrust America into a full-scale revolution. The White House was shaken and I was willing to help shake that administration at any level and at any personal cost.

In May of 1969, I arrived back at my home in California. Seeing my wife and family was the only thing I looked forward to. Thirty days later, I reported to my next duty station, Hamilton Air Base in California. I was told about the last munitions airman who had arrived from Vietnam a few weeks before me. He had been cross-trained into the security police, like me.

He had threatened to destroy Hamilton Air Force Base by detonating a nuclear warhead missile on an aircraft. His request was simple,

"Discharge me today or I'll blow this place to pieces."

He disclosed classified information about nuclear weapons, causing a wave of media attention and protestors. He eventually surrendered and was taken to a nearby military psychiatric ward. Little did I know that my life was about to follow a similar route.

I was the second munitions person to arrive from Vietnam. Meanwhile, my oldest brother, Fred, was working at a military contract company. Many top brass officers from air force bases around the country, including the Hamilton AFB, were under congressional investigation for the embezzlement of millions of dollars. The Vietnam War was making some officers rich.

Fred instructed me to write down everything that concerned me and to ask anyone else who felt they were being mistreated by the military. I proceeded to do this. I began by questioning other airmen about their military welfare. I assured them that my brother and his contacts in Washington, DC, would get all this information about this unjust, brutal war to the right people.

I proceeded through the chain of command, presenting all the collected questions and complaints to the commander of the security police squadron. Perhaps this was to be my open ticket for discharge. As I continued through the military chain of command, I spoke to the base chaplain and the military psychiatrist. I told them my side of the story. I also told them about my early childhood mystical experiences, because I wanted them to know more about who I was. I spoke for twenty minutes with the psychiatrist—less with the chaplain—during which time I felt a familiar force intuitively leading me somewhere. As I walked back to the security police squadron, I knew somehow that I would never walk through those hallways again.

My first sergeant told me to report to the psychiatrist immediately. I complied. I told the psychiatrist that I wanted to know more about the teaching of the Rosicrucian Order. I felt sure they would be able to answer some of my questions concerning the great white light with powers beyond human reason and comprehension. In retrospect, I know they must have considered this kind of talk completely irrational; I remember feeling certain that day would be my last on Hamilton Air Force Base.

The doctor and a lieutenant colonel and two other airmen told me they plan to take me to Travis Air Base for further testing. I agreed unhesitatingly. I asked if I could go home and get my toothbrush and explain my situation to my wife. They denied this request, saying I was to leave for the base immediately. I asked politely if I could make a couple of phone calls.

This they permitted. I called Fred. He advised me to go along with it and said there was a lot more going on than could be explained at that time. I called my wife at work and told her I would be unable to come home. I informed her briefly of what was happening. Then I told her to call Fred. She did not understand. How could she?

I was placed in a secured vehicle so that I could not escape; this seemed very strange. I remained silent and upset—upset about being separated from my wife and upset for having been apprehended and escorted to some unknown destination in a locked vehicle. The unjustness of the war was raging within me, and I began to write down some notes.

The driver asked what and why I was writing. I replied, "What would you say if you were me, and who would you tell?" He and the other guard fell silent. Obviously, they were not allowed to talk with me. I finally told them I was writing to Senator Cranston. I then lapsed back into silence and more writing. Before I had been securely locked in the vehicle, I remembered asking the doctor and the colonel, "How long will I be gone? What kind of testing?"

I was told I would be gone for only a couple of days, and this is what I had told my wife.

While recording these things, I began to think back to my last days on the Hamilton Air Force Base. At one point, the military had offered me my old career field back, since my military record was flawless in every respect. I had been offered the highest plum in my chosen field—Explosives Ordinance Disposal, or nuclear weapons school. I had enlisted and was obligated to remain in the US Air Force for four years. I refused their offer because after my one year tour of duty in Viet-Nam everything had changed unexpectedly. However, if I had chosen to finish my obligation I could have been discharged like everyone else, or perhaps continued a military career. I thought about how much easier that might have been. I could have finished my pilot training at the base aero club. I thought about my wife, alone, confused, emotionally upset. Why should she suffer? My eyes glazed over but I held the tears back.

The previous two years rolled through my mind as we drove. Then, a particular incident came clearly into focus. I recalled entering the security police squadron one morning a few months before. I had a smile on my face and was feeling particularly cheerful. I was dressed like a soldier—proud, sharp, and willing. I was ready to carry out my role in the military with enthusiasm. As I entered the first sergeant's office, we exchanged morning greetings in a gesture of proper military etiquette. I deposited fifteen cents in a small bowl for a cup of coffee. Noticing that several cups next to the coffee pot were dirty, I chose the dirtiest cup in which the coffee had been left to harden over the weekend, and went to the nearby sink to wash it. It was a small gesture, but one of goodwill, nonetheless. Returning and filling this now-clean cup with black coffee, I proceeded to my office. I was in charge of teaching several new airmen the first series of tests on becoming a security policeman. I felt sharp, and I was. I was in harmony with the military.

While preparing the class, I had a call over the intercom system from the first sergeant. He told me kindly that I had taken the major's cup. I checked the cup to see if the major's name was etched on it somewhere, but I found nothing. Politely I told the first sergeant that I had washed an old chow-hall cup, and that, I assured him, it was not the commander's. I returned to my work. Almost immediately the sergeant called again. By now everyone in my office was listening intently to the conversation. They were curious. Once again he repeated his accusation—I had the major's cup.

"No, sir," I replied firmly. "You are mistaken."

Seconds later the sergeant interrupted again. "The major is coming to get his cup," he said in a harsh voice. I slammed my hands, palms-down on my desk, shocking everyone. Then, grabbing my hot cup of coffee and clenching it like a grenade, I storm-trooped up to his office. Coffee spilled over my hand and onto my pant legs without my noticing. I met the major halfway up the stairs leading to his office. I threw the cup at him, scorching him with my anger as well as the hot coffee. The cup flew past him, smashing like an explosion into dozens of pieces attempting to destroy his inflated military ego.

The power of my anger filled the room with an explosive discharge of words. In a rage and feeling as if some unknown person were speaking through me, I said, "I am in command here, Major! I am in command of my life! I am in command of who I am and of where I am traveling. To you and this establishment, I am no more!"

He stood there, stunned for a moment. Then he screamed something about insubordination or some kind of military bafflegab. His voice faded as I raced from the building. In my furious rage, I swore that I would reduce this base and all military edifices into a dust invisible to the naked eye. Strangely, the words seemed to set me free.

I drove home and told Patsy that I had quit my job. She only smiled and made love to me. Later, I explained to her that while I had once loved and had the greatest respect for the military, I had now rediscovered myself, and my only desire was to be free. Moreover, I wanted to free others who were also held in bondage by military personnel; their lives were, likewise, suspended in false illusions. I felt that God was free and His law was supreme over all. I felt the cause of freedom was trapped by manufactured military, governmental and political illusions. Freedom was gained by the quest of truth, I thought. I knew then that I had to fight my own way to freedom.

I needed to discover an absolute meaning of freedom; it had become essential for my existence. And it would be the love of truth that would ultimately set me free. Truth indeed was becoming the primary quest in my life, which would result in the discovery of an enlightened freedom. At that time, I could not quite understand it, but afterwards, it would become clearer. And, years later, I was led to another personal quest—that of wisdom.

Then I recalled another officer with whom I had spoken around the same time. He was in charge of the security police squadron. He was kind and understanding. He asked me if I wanted out of the military. I told him yes and why. I told him I felt some kind of force pointing me to a new sense of direction and above all, a profound sense of freedom; I felt that force to be my personal and true path to victory. I felt compelled to follow its direction. He agreed that I should follow my inner convictions.

Suddenly, I snapped out of my reverie and returned to my present circumstances—a captive in the secured military vehicle. I became frightened and paranoid, realizing I was heading into an unknown situation. My biggest fear was that I had no control over these circumstances or my destiny. I gazed outside the window from my backseat position and pressed my forehead against the glass. For a moment, my mind went blank as I numbly watched other vehicles pass—perhaps into their own uncharted destinies. Finally, I let go of the whole situation. I, too, was destined for a new experience and an unknown destination. But at the moment, I was being held captive; I had become the enemy.

Suddenly, I burst into uncontrollable laughter. The drivers began to laugh with me, although they did not know why. It was simply contagious. They asked me what was so funny.

"Me, you, this situation, the military, nothing and everything, God knows what," I replied, and began to tell them all about my last days at Hamilton Air Force Base. I began talking to them like old military buddies. They responded in kind, enjoying my stories.

I told them about the time, after the weekly police inspections, that I had been in charge of selecting a movie that would boost the morale of the security policemen. After the inspection, the officer in charge and all the other men anxiously gathered for the movie. There were about sixty men in all. I clicked the projector on, and an old black and white film began to scratch through. The title appeared, The Industrial Revolution of the USSR. Out of the silence, the officer in charge whispered to the first lieutenant, then left and went up the stairs. I suppose he did not like the movie, or perhaps he had seen it before.

I overheard the first lieutenant whisper to my supervisor, "Is there something wrong with this movie?" My supervisor had no idea. Then the lieutenant and my supervisor approached me. I was asked why I had chosen that particular movie, and what it had to do with the security police. I told him there must have been some kind of mistake and asked if I should shut it off.

But the lieutenant said, "No, it will be over in just a few minutes. But Sergeant Weiss, next time let's see the assigned movies."

Another week passed and it was another Monday morning inspection along with another scheduled film to start the week. Some of my Mexican friends in the training room had a difficult time holding their laughter during the inspection. By this time, word had gotten out. More than half the men knew there was another surprise movie in store.

The lights went out and a few seconds passed. Then the movie began with a burst of laughter by hundreds of US servicemen. The theme song of the well-known entertainer, Bob Hope, began to play. Uproarious laughter filled the room. Some airmen began cheering as the Bob Hope theme song captured the audience of the security policemen. My Mexican friends finally let loose, and soon the entire squadron was choking with laughter. The first lieutenant walked toward me at the back of the room and said, "I'm glad to see you have a sense of humor, Sergeant Weiss, but I think I will pick up next week's movie. Please see me in my office as soon as possible." I agreed, as the movie entitled, Bob Hope in Vietnam, continued to roll.

As I recounted the story, the airmen in the vehicle laughed along with me. By this time we were approaching the Travis Air Base in California. Two men in white coats escorted me to a building called the Ward F-1 Psychiatric Department. While I waited in the hall for a few minutes, I noticed everyone was dressed in blue pajamas and white slippers. I saw another man at the end of the hall wandering around in a straitjacket. Yet another man approached me, saluted in the fashion of a Roman legionnaire, and said, "Hail, Caesar!"

I went over to the drinking fountain to get a sip of water. A tall black man approached me and warned, "I don't know why you're here, but if you want to get out, don't take any of these fuckin' drugs they give you."

That freaked me out. I was tapped on the shoulder and introduced to a military doctor. At his request, I followed him to his office. After asking me a few general questions, he fell silent. I was silent for a moment, and then asked him how long I would be there.

He said, "Until you know why you are the way you are."

"That could take a thousand years," I replied seriously.

"Then a thousand years it will be."

He went on to ask me a few psychiatric questions such as, "Why can't you fuck an alligator?" I told him I had never desired to fuck an alligator, and asked why I was being asked such a weird question. He said that would be all for now and that I would be taught the routine of living in the ward. I was shown to my bunk bed. I surrendered my military clothes in exchange for blue pajamas and white slippers. All my personal things were confiscated, including the information I had been writing down. I was told that dinner would be served shortly, and that I must eat alone in my room. A short time after dinner, I was served a group of tiny pills in a small paper cup alongside of which was another cup filled with water. I swallowed it all without protest. It relaxed me, and gradually I drifted into a deep sleep. My last thoughts were of my wife, Patsy. I missed her so much; I needed her with me.

I was awakened by an intercom system that blared, "Good morning. It's time to get up." Just then I recalled a dream I'd had about green caterpillars that were stuck all over my body like blood-sucking leaches. My eyes opened but I couldn't move. A staff member of the ward approached me and helped me out of my bed. I stood like a frozen man; frozen, but with every muscle in my body trembling. I was led out of my room like a zombie. I could sense my lips were drooping, as though I had taken an overdose of Novocain. I remembered ruefully the drugs I had been administered the evening before.

As I attempted to walk, I felt one foot slide in front of the other without any coordination. My breath was shallow. Was I Dr. Frankenstein's experimental monster? My mind blurred; I simply stood in the hallway, staring into endless space. Finally, I moved my left index finger. I waved my left hand in a feeble gesture to say hello to my new black friend, David. We sat down together, and he warned me again not to take those drugs or I would go crazy. He comforted me and explained the ward system. He said, "This is the military's cuckoo's nest, and we got to fly out of here."

I told him I felt like what a frozen cucumber must feel like.

"Well, you should have seen me after a few days on those damn drugs," he said wryly. I understood what he meant. "You just thaw out, man," he said, "and I'll show you exactly what's going on and what you should do."

A few days passed and I began to learn the ward rules. There were six levels. You were automatically assigned to level one when you first arrived, which meant you were confined to your room and had to ask permission to even go to the bathroom. Breakfast, lunch and dinner served at your bedside on a tray. Everyone participated in daily sessions of group therapy. You could pass through the ranks of the six levels only by a unanimous vote from the group. In level two, you could leave your room and eat together with others in the ward. The bathroom privileges were returned as well. In level three, you could eat at the chow hall with your friends. Level four allowed you to participate in outdoor sports, swimming, making ceramics, leather works, and other creative activities. Once at level five, you could go anywhere on the base with a curfew time of midnight. Level six meant you could leave the base for a weekend as long as you had an adult sponsor or an immediate family member to sign you out.

Later that week I was led to a large room where I met with the hospital staff. I was instructed to sit in the center of the room, where a chair was placed for me. I was hit with a barrage of questions from the various staff members and was closely scrutinized as I responded. I was then escorted back to my room.

During the first week, a captain—a woman—confronted me. In silence, she walked me back to her office and then she spoke, her voice growing increasingly louder. She was testing my anger, I thought, so I stood like a soldier under discipline. Five inches from my face, she drilled me like a sergeant. I gave her my attention and respect—God knows what would have happened to me had I reacted adversely. She asked arrogantly, "What makes you so certain that your brother has any influential contact in Congress?"

"If my brother doesn't have this contact in Congress," I answered placidly, "then I'm living in the wrong country." Calmly, I stated that all I was reporting would be hand-delivered to Senator Cranston, along with other many incriminating items that were being collected to be read on the Senate floor. I reminded her that Senator Cranston was against the Vietnam War as much as I was, and that he would personally call and speak to the hospital staff to determine if my civil rights had been violated.

I was anxious to see my wife. In three short weeks and with the group's favor, I earned my way to a level six. David had been right about everything, and I admired him for embracing me as a friend. I was kindly warned by others who said to be careful—that home might not be the way it all used to be. I was somewhat haunted by their message, but Patsy meant far too much to me to ever consider being away from her again. Nonetheless, I left feeling positive and was anxious to get home. I was thrilled that now I could leave the walls of confinement and once again be with Patsy. My brother, David, drove up in his car to sign me out. His wife, Susan, drove in my car for my return trip home. But, just as I began thinking that things were returning to normal, I felt a wave of confused sadness permeate my whole body.

David and Susan drove back home together and I continued on my way to Patsy's home. On the two-hour drive, my sadness passed and I became excited about reuniting with her and having love in my life again. But I became silent as I approached the house, and a feeling of terror constricted my throat. Again I felt crushed by a deep emotional sadness. I could scarcely believe that I suddenly felt her love for me was vanishing from my life. I was bewildered: was her love for me somehow disappearing, or was my love for her disappearing from my heart? Why was I feeling this way? Could this be true? How could I even know this? I was certain that I was really beginning to crack up. Somehow I knew it was the end.

I stopped by the roadside to compose myself, trying to shake it off as a bad dream. Then, filled with trepidation, I hurried off to our house. I parked in the driveway, waiting in my car with my hands and head on the steering wheel as tears streamed down my face. I heard the door open and close as she got into my car. I looked at her, softly imploring, "Please don't leave me. I love you and I need you more than ever."

Silence ensued. Patsy revealed that she had been confused for a long time; and could endure no more. I sat there, stunned as she informed me that she had filed for a divorce. She explained, through tears, that she could not love me anymore because of all the emotional confusion the military had caused in our lives. With a quiet goodbye, she closed the car door and left. It was that quick. I was left in a jumbled state of shock, panic, a blitzkreig of pain and sadness.

Patsy had chosen to leave me. I was numb. I drove to my parents' house and stared into a void of deep sadness for the duration of my weekend of freedom.

When I returned to the ward, I was placed into a group where each level-six person was required to explain all the events and circumstances of their leave. The group would then question the person "on stand" to determine any level changes. I explained to them that my wife had filed for a divorce. I told the group that I was thankful for all their support and love. No one had any questions as I wept like a child in front of them.

A few weeks later, while I was playing Frisbee in the ward hallway, the doctor called me into his office. Senator Cranston was on the phone, he said, and was asking about the injuries I had received while in Vietnam. After I told him, the doctor thanked me and said he would speak to me later.

A swift judgment was returned by the hospital staff. It read: "If any patient had any military misconduct, such a person would be discharged accordingly. However, to all other patients, like myself [Senator Cranston], an honorable discharge would be granted, or the right to resume their prior job and rank in the service."

We were given the choice. Many, like me, wanted out. I retained military honors and retired with privileges and an honorable discharge.

Once again, I was free. This time, freedom meant I was at liberty to look directly into the face of pain and confusion and the deep, unexplained aspects of human life. My cherished ideals of marriage and the patriotism I had felt for my country had been irrevocably twisted inside my heart. I remembered the twin loves I had espoused as I left for Vietnam. I had now been stripped of all that I had held dear—love for my woman and love for my country. *But I was free.*

Chapter Four

❧ THE UNKNOWN SPEAKER ☙

People tell me it's a sin to know and feel too much within.
—A Simple Twist of Fate, Bob Dylan

ALTHOUGH I MADE some feeble attempts at reconciliation, I failed to resolve our marriage issues and was quickly served my divorce papers. Patsy returned to live in her small town to care for her drunken father. I later learned that my Dad was offended enough to remove him permanently from his pathetic drunken life. The alcoholic dad had told my father the only reason Patsy had married me was that she would receive my beneficiary money of ten thousand dollars if I was killed in Vietnam. My mother, in the meantime, had warned her not to divorce me until I had been back from Vietnam. This was my mother's attempt to protect me from almost certain self-destruction. God knows how I would had reacted had she left me earlier. The war and military life had created a more fragile me than I had ever known.

After my discharge from the military, I moved in with my oldest brother, Fred. His marriage was also on the rocks and we shared many common bonds and interests. During this time, I tried to pick up the scattered emotional pieces of my life. We later enrolled in a college together. I was ripe for the attraction of all manner of leftist radical movements and plunged right in. I became actively involved with the leaders of revolutionary activists. I learned much about the radical communist and socialist ideas of the time—especially about their politics and anti-war movements. Vietnam, in particular, was a sore spot with me. Altogether, these new ideas struck my heart deeply.

I still cannot entirely explain why at that period of my life I felt compelled to spread my revolutionary ideas in Sweden—I wanted desperately to go there. Certainly, there were people who strongly influenced me. During that period, I became reacquainted with an old friend, Peter, from high school. He shared my passion. We collected two thousand dollars for our trip in one afternoon. His old girlfriend, who had been awarded $10,000 from the military when her husband was killed in Vietnam, gave Peter a thousand dollars. We went together to meet her at a nearby restaurant, but we invented a story about my being his bodyguard, who had to get him out of the country as soon as possible. I sat at another booth while they negotiated the deal. She handed him the cash. I rose from my seat, nodded at them as I walked past, and went to wait for Peter in my car. When Peter joined me, we took off in triumphant laughter.

On our way across town, we picked up two female hitchhikers. One of them told me she needed a good VW bug for school and asked if I knew of any for sale. That afternoon I sold her my car for a thousand dollars, and, lo and behold—within twenty-four hours we were booked on Icelandic Airlines, going one way to Oslo, Norway. I began to relax, although a deep emptiness and loneliness was still rooted in my heart.

Upon our arrival, I felt an odd familiarity with the Nordic people and their land. But Peter and I decided to go our separate ways after Denmark. I took a short visit to Hamburg, Germany, and then went to Stockholm, Sweden, where I remained a few months. I had only twelve dollars remaining when I arrived, so I paid for a room for one night and spent the rest of my money on a fish dinner. The next morning, I walked one block from the place where I was staying, turned right, and walked up another block. I found myself in front of a sign written in English that read, "American Deserters' Committee."

As I was contemplating whether to check this program out, a Swedish policeman approached me, asking politely to state my business. He wanted to search me, and I agreed without hesitation. I then passed by the guard to speak with the committee upstairs. I spoke to them briefly about my reasons for leaving the United States. They offered to find me a place to live with some Swedish students and enroll me into the University of Stockholm to learn the Swedish language. However, I was not eligible for any income because I was not an American deserter and I was not traveling without a passport. This group's policy was to give all American deserters a place to stay, an income, and immediate enrollment in school. I was on my way to becoming a citizen of this country. I felt love from a country I had previously known nothing of, and in return, I fell in love with Sweden. I cherished it as a rare jewel, a crown of freedom, although deep inside I remained barren, as though God had abandoned his creation and left it to feed on its own dung and ignorance. At least I still had my income from the US government.

After two months, the communist influence I had become acquainted with in college began to grow stronger. I participated in Swedish anti-war marches, protesting the American involvement in Vietnam.

After a time, my involvement with the American Deserters' Committee began to change. In fact, it changed abruptly when one day I uttered a few words concerning revolutionaries. It happened during the time I was planning to visit Moscow. I wanted to meet the Russian people and understand more about the Soviet way of life. I had received an offer from some Soviet officials who frequented the ADC.

Would I be interested in working in North Vietnam as a weapons and intelligence operator? But, during one of the ADC's regular meetings, I said in a stern voice that captured everyone's attention, "This revolution has nothing to offer in your conquered place—nothing except political and tyrannical control."

I was bombarded with a barrage of questions: "Why are you here on a visa and passport?" "Why do you receive money from the US government?" Suddenly, it seemed as though I had encountered a new adversary. Once again, I felt the meaning of being an enemy and realized that I had ricocheted from one monstrous political ideology into another. Instantly and without warning, I slammed their revolutionary communistic cause and threw it back into their ideological faces. From out of nowhere these words fell from my lips, as if a voice of truth had split my present mindset into pieces. I went into a rage, stating that I could no longer advocate or support their revolutionary concepts of Marxist philosophy. This was not my path to freedom or a means to grasp the roots of wisdom. I told them that my quest for freedom was a far greater plan than their machinations of political control and revolution. However, a burning hatred remained towards the US involvement and escalation of the Viet Nam war.

Needless to say, my trip to Moscow was placed permanently on hold; my invitation by the Soviets to work in North Vietnam would have to wait. I simply felt the lack of freedom and wisdom. I needed to voice my opposition. I became aware that God was a power with a great plan. For people who value freedom and yearn to be free, the Creator's laws can awaken in them a new hope.

"For me, I like to look back at that profound message and believe I attempted to overcome the ways of revolution, wars, sickness, and even death." This I believed to be our true enemy—our true adversary is the serpent that consumes itself and perpetuates its cause and likeness. At least to me, this I believe to be an absolute truth.

I made great friends with the Swedish people, though I was no longer so popular with my own fellow citizens in the ADC. I felt isolated again. Once again, I began desperately searching for a way to fill this emptiness—this void where I believed only God could show His greater light of knowledge. I called my parents and brother and informed them of my whereabouts. Fred suggested I return home and move to Montana. He was planning to build a fish hatchery there and wanted me to join him. I felt deserted and missed my family and decided to give the project a try.

I boarded Icelandic Airlines, wearing a double-breasted green suit and tie that I had custom made in Singapore during my leave from Phu Cat. I sat and made myself comfortable on the plane while slowly removing the assorted communist buttons I wore on my lapel. I remained a self-contradiction and a stranger in a strange land. But worse than that, it would be some time until I would fully realize the meaning of my own new and personal revolutionary thoughts. I had a lot to think about.

I sat next to a man. He introduced himself and said he was from Czechoslovakia. He asked if I had ever seen a check bounce? My reply was no. He responded by bouncing up on the seat of the plane. His sense of humor and ease of conversation allowed me to put my thoughts on temporary hold. The plane landed in Keflavik, Iceland, and I went on a walking tour to the extraordinary capital city of Reykjavik. I had a sudden but lasting and a familiar premonition that someday I would return to Iceland.

Arriving at Kennedy Airport in New York, I was the last to step off the plane. The Icelandic stewardess wished me the best of luck and suggested that I return to visit Iceland soon. I did not know at the time how important Iceland would become in my life.

As I stepped off the plane, two men wearing dark suits approached me. One of the men showed me a list of names. After I pointed at my name and showed my identification, I was immediately escorted to a private room. I was questioned by two more men and asked several questions concerning my time and activities and the people I associated with in Stockholm, in particular. I was searched from head to toe. I had several books from the Soviet embassy and all kinds of communist literature from the Swedish students, including the famous Red Books by Chairman Mao of the People's Republic of China. Eventually I was released. I placed a call to my brother, Dave, asking him to wire me some money for my plane ticket home. I never knew who those men were, but clearly I had been monitored, followed, questioned and searched.

I have always wondered why. And why was I questioned about my actions later. I must have been a perceived threat, but to whom and for what? I never considered myself a threat, but perhaps I was perceived as such to those who followed the leftist movement and the party members of the ADC. I knew of no other reason why I had attracted their attention. I assumed those men were government officials and I had no idea how they knew I was arriving, let alone how I got on their list. I figured it was some kind of US intelligence investigative department concerned about my involvement with the leftist philosophy. I never asked them for any identification nor do I recall them showing me any. It is highly likely they were CIA operatives.

Like a stray cat sensitive and alert to the present moment, I was creeping through a mysterious jungle—my fate unknown. My strength was my desire to be free. I was spellbound by freedom in a world that was enslaved to a number of lies that had been promulgated for generations. Freedom was my sword to salvation. Without it, I felt doomed by the world's dictates. I needed the freedom to know myself and the origin and destiny of my human heritage. But was I a threat simply because of my quest for freedom?

It was not until sometime later that I discovered the true meaning of freedom—when I posed the question in a solemn prayer to the God of my heart: What have we come to do in the temple of Freedom? It was freedom I desired, along with a vision of freedom's purpose and the power to transform that vision into reality. This is of critical importance for those who want to remain free.

Years later, I began to understand the profound connection between my destiny and America's as it relates to all humankind. This value, along with my later encounters at Mt. Shasta City, California was a major pivotal point that changed the course of my attitude and the direction of my life, forever. My experiences on the mountain itself validated these words I had spewed out on that fateful day at the ADC in front of the Soviets who presided over the meeting. I quickly grew to recognize that each country and each individual who love freedom must fight and win that battle to preserve a sacred destiny. This will occur when the nations are willing to listen, remember and honor freedom's inner voice of the unknown speaker. The unknown speaker is the immortal voice that speaks from the fountainhead of freedom to all who dare to listen.

Chapter Five

A New Path

At the touch of love, everyone becomes a poet.

—Plato

I HOVERED OVER PILES of books and the laptop computer on one end of the large oval dining room table while Wanda worked silently at the other end. I was still in my navy blue terry bathrobe. I stared at the computer screen, thinking, my hands poised over the keys. With her dark hair in pin curls and dressed in a faded Denver Broncos sweatshirt and Levis, Wanda gulped coffee and read through her red horn-rimmed glasses.

"That stuff will kill you, Wanda." Somehow I did not need to turn my eyes from the screen to know she was holding the ever-present coffee cup. "You won't catch me drinking that poison anymore."

"Get this!" Wanda hooted, ignoring me while she ran her fingers down the page of the fat volume—a spiral-bound publication with a green cardstock cover and large black letters on the front: *The Hollow Globe*. It was written by Professor Wm. F. Lyon and published in 1871, through the resources of M.L. Sherman, MD.

" 'You're the man that I have been searching after,' " she read aloud, assuming a thick British accent. " 'The very man I was to find, and we have a large amount of business that we must transact together, but I am not fully prepared to state the nature of that business, for I do not seem to understand it myself.' "

Wanda removed her glasses and looked for my reaction. My concentration was already broken, although my eyes remained on the screen, fingers still poised to type. "Right, where have we heard that before?"

"Amway!" was our quick, simultaneous and jovial response. Oh yes, nearly everyone has been approached and remembered the well-known American multi-level marketing business that required you to participate in a secret meeting with a group of people before they would divulge what it was all about.

After a time, Wanda looked up. "But this is really amazing, Dan. I mean, this Dr. Sherman knew who he needed. No question. He was actually led to Lyon to do the writing for him."

"That's what I've been trying to tell you, Wanda," I said, shaking my head. "This kind of thing has been going on all the time—from the very beginning. See, I'm not the only one who's obsessed with freedom and truth. When you know what your mission is on this Earth, you are led to the right people and places and events. You are literally given everything you need to do the job. That's why you moved from Denver to next door to me—to help. Since Diane walked out of here with half my tapes and notes, I've been swimming in chaos. I can't do it alone. You were literally a gift placed at my doorstep."

"Come on, that's a little much, Dan."

"No, it isn't. God gives us what we need when we need it. He always has. That book you're reading—it was written in—when?"

She flipped back to the previous page. "September, 1868, it says here."

"See what I mean? And it goes back a lot further than that for me, remember. Wanda, I'm not just some drugged-out hippie still suffering from 'Nam jungle shock, believe me."

I got up and shuffled around the table, my arms circling in the air as I spoke. I emphasized every other word in husky punches of breath, exhausted from telling the same story over and over. "Well, maybe I am that, but I'm not crazy. This theory goes way back to the beginning of time. Every single culture has some kind of reference to the Hollow Earth in their teachings—the Bible, even! It's all there, Wanda. Look, Paradise, Valhalla, Asgarth, Shangri-La, and the Hopi Indians—everywhere. It's there—it exists, Wanda. Even the Koran proclaims it! I know it. I can feel it. I feel some inner connection with the entire Nordic area. I've been drawn there for a long time. I just want to get there and see it for myself. That's our home—Paradise—the Garden of Eden. The Golden Age is where we came from."

I breathed a long sigh of mutual understanding. Just two simple human beings adrift at sea, their vessel a large oval dining room table covered with documents, fighting off an earthly storm that would feel no remorse in taking them both out, down into the icy depths without a trace, were it not for providence, sweet providence. We sighed. We understood for a brief second. Then it was lost. We released weak smiles. What were we getting into?

"Wanda," I said, speaking calmly now, "I've lived, breathed, dreamed, been beamed up, lectured to, burned out, tossed around—heck, I've literally been drowning in this stuff practically all my life! When is it all going to end? I'm going under with it, if I'm not careful." My voice must have belied the look on my face. I felt the heat of frustration emanating from my body. "I need help, Wanda!"

"Yeah, well, if I get up to my neck in shit with you, we'll never get anywhere, Danny boy. If you want me to help you, this thing has to be focused, honed, hand-crafted like a spear. Sharpened, bound, and—"

"Yeah, that's it! A spear that we can aim, tosses, and hit the target, right in the center of the Earth. Bingo! Let's do it, Wanda. It's time. I'm ready to go."

"I may be beginning to get the picture." Wanda was not really happy about this, but resolute. "Give me everything; I'll get it all on tape—when it starts, how it feels, where you think you're going with it, how all this other stuff fits in—everything! Your journal notes, files, whatever you can remember. Let's do it."

"Okay, here." I handed Wanda a thick file folder. "Start with this."

During the early 1970s, I was deeply concerned with finding my other half, a woman who could share with me all my inner feelings, thoughts, and emotions. I began to write poems. The pencil seemed to be both a magic wand and a sword that would cut its way back into me. I discovered a pattern in my writing that provided a bridge from a bitter and distraught past across to a new understanding.

My writing took on various meanings, suggestions about my inner self that lifted me from many destructive ideas. I began to reform my thinking by writing poems about love that seemed to come from an internal source. A great change was growing within me, and I began to make psychic predictions in my poetry. My poems helped me to transcend the ghosts of the past. Poetry became a tool through which my life unfolded.

Certain women with whom I came into contact moved me, metaphysically. I wrote a poem to each. In the brief moments that I shared with each woman, I could envision a new world yet unrealized. Each vision was a reflection of an inner world, allowing me to glimpse into my past, present, and future. As I stood still within the present, the past and future radiated like swirling shadows behind and before me. However, I was seeking that certain someone who could fully allow me to see and embrace my inner being.

This pathway of poetry took a prominent place in my life. I not only wanted to write about what I had observed, but I wanted to release the intuitive feelings, thoughts, and power of the eternal that I sensed and witnessed by the reflection of each woman I encountered. My need for a woman to share my love with grew stronger and deeper as I continued to grow within myself. With my poetry, I attempted to interpret and contact the source of the eternal. My love was growing in a universal sense as a deep personal relationship with God. A mysterious cloud or a mist surrounded certain women while I stood in their presence, and I knew I would eventually find that mirror reflection. She would be a key to lead me to the meaning and purpose of my cosmic origin and destiny. From that point, I knew I would be free to proceed with the vision that had come from within. It was a glimpse of the eternal and I knew that someday I would totally realize myself with it.

The year was 1971. The fish hatchery never materialized. My brother, Fred, had his own problems, so we enrolled in college together. I was deeply concerned about the Earth's environment, the US government, and political leadership throughout the world. Over time, I began to lose contact with my writing because I was attempting to explain with words what simply could not be articulated. My emotions and desires were united, becoming a primary direction in my life. I knew I was coming closer to finding that one woman who could reflect all that was within me: my greater being, the being I had experienced as a child. I yearned to express my deepest feelings. I put aside my path of poetry and began to search for a different key to free my inner thoughts.

During a militant political rally which I had organized over the Easter holiday, I met a special woman who caused significant and rapid changes to occur in my life. This became a true turning point that ultimately allowed me to receive a glimpse of where I had come from. Her name was Katrina. When I first set eyes on her, I immediately perceived a cloudlike mist that settled in the center of my mind. The vision I saw was a reflection of my inner self. It allowed me to reach into an endless eternity. I felt whole and at the same time incomplete, but I had some understanding of my deepest feelings of incompleteness. I felt light in weight. I somehow knew I could be honest with this woman and share all my deepest feelings. Her beauty was like an invitation to Paradise. A quivering vibration rushed throughout my entire body. I could see many faces from different times and of different races of people. I knew at once we were going to be together for some time.

It was the eve of Good Friday, but felt more like Passover. On the evening of April 8, 1971, Katrina, my friend, Ed, and I went on a political escapade. Our mission was to erect a giant stone pillar about ten feet tall in the center of a meadow as a protest against a by-pass project planned for the meadow. It signified that we would not be moved from our cause. We borrowed a pickup truck, and I rode in the back on top of the stone pillar while Katrina and Ed rode in front. My thoughts were tangled as we raced silently to the top of the hill. The pillar became symbolic of a cross to me. Internally, I began to receive the knowledge of Jesus' crucifixion growing within me. I believed I was being brought up to the hill where Jesus had been crucified. While our pillar was just symbolic, it was somehow also real, as though I had received a vision and memory of the time of the crucifixion.

We dug a hole, placed the pillar inside, and cemented it in place. Next, we placed a kerosene canister on top and lit it so that the pillar glowed mysteriously across the vacant field. At that moment I discovered that I had lost my ring.

The ring had been broken in the middle. To me, it symbolized a mystery concerning the North Pole. The broken ring symbolized a secret entrance into the Earth's interior by way of the North Pole. This was a mystery that had been revealed to me, intuitively. I had purchased this

*Jesus replied to the thief at his side,
"Truly I say to you, Today shall you be with me in paradise."*

ring in a pawnshop along with an authentic Nazi flag and an old sword. One day, in a symbolic quest for power, I draped the flag around me, drew the sword and walked through town on my way home. With my imagination running ahead of me, I felt a cold chill from a long-hidden memory about my display of fantasy. The ring symbolized a great hidden world and the broken space symbolized the secret entrance into that world. The sword was my quest for justice, power, and world leadership. The flag haunted me with strange thoughts that somehow seemed like real memories.

While in college, I had many dreams in German. I rediscovered old acquaintances who were all clad in Nazi uniforms. Even as a child of ten, I remember draping a white cloth over the kitchen table and carefully constructing an iron cross with black and red fabric. When it was completed, I sat and marveled at it. I proudly waited for my brother and parents to come home to compliment my handiwork. To my surprise, I was bombarded with the words that stung, "Where in hell did you get that idea?" Ironically, my brother, David, an ardent college student of the Russian language, had an old authentic USSR flag displayed on his bedroom wall.

I had been brought up through the ranks of the Catholic Church until my Mother a devout Sicilian Catholic, yanked me away from the demanding and homosexual priests. During Thursday afternoons my fifth grade class mates who waited outside the school yard always shuttered in fear wondering if they would ever get to come home while being hauled off to catechism. I remained in a deep silence while Nuns pressured me to recite words beyond my ability to speak. My Mother set me free by putting the priests in their place by scorching them with flaming words that severed me from the church forever. She had the last word saying that Dan will find God in his own way.

"I thought you would be happy to see this!" I responded, running to my bedroom. I cried, curled up with my dog, and eventually fell asleep. I was never questioned about my emotional outburst or the flag I left on the kitchen table.

In the course of our mission, Katrina's face was glowing like the full moon. I began to receive loving impressions from her spirit—they seemed to radiate throughout my entire being. Her love surrounded me with gentle warmth. The feeling accelerated. Her soul was a reflection from the light of the full moon, yet she seemed to have touched the light of sun within me. Her moonlike force pulled me into the memory of a deep emotional experience of love. A love filled with peace, joy, and a sense of a very ancient unknown karmic destiny. I soared into the sun's light like an arrow. The moon became a living substance and melted before my eyes, turning into a liquid that poured into my heart and condensed into a vaporous cloud before me. At that moment the moon was an illusion; it was no more. I felt subtle warmth. I was free and traveling like a swift arrow, following a path directly toward an inner sun within me. This was the encounter and our unity as brother sun and sister moon.

After we had completed our symbol of protest against the meadow by-pass, we drove back to Ed's. It was close to midnight. We heard rumors from friends that some radical groups had bombed the Bank of America that same evening. Their radical plans seemed, coincidentally, to parallel with our own, so we decided to abandon our other ideas for fear of being pinned to the bombing. Peter, another friend, was to climb up the flagpole at the city center and hoist our own flag—it was yellow with a blue, five-pointed star in the middle. This signified our political protest against the Vietnam War. Our other comrades were to have stolen large sticks of TNT to be placed in the drain system underneath the new highway construction.

Another friend, Pat, said the explosives at the quarry had been stolen before he got there, indicating that someone else had taken them while we set ourselves up for the blame. All was not going as planned. I suggested that we head out of town for a while to let things cool.

The local newspaper picked up on it with the headlines: "What If They Gave a Party and Nobody Came," and "The 'Party' Plans One in the Meadow." It was not until later that the FBI began to question my family and friends concerning the destruction of the banks in our community.

"Let's go to Montana," I suggested.

Ed was against it, so he dropped Katrina and me off in town. We picked up my car, drove to a nearby church, and parked for a short time. Within minutes, Katrina and I changed our plans to Montana and headed east to Reno, Nevada, to get married. We drove to my parents' home first and left a note on the table explaining our plans and that if anyone should ask about the Bank of America bombing, that I had nothing to do with any such action. Katrina and I drove to Watsonville to pick up some of her clothes. She left a message for her mother and told her not to worry and that we were on our way to be married, and would return by Easter morning.

We left with a small amount of money and a Shell gasoline credit card that served us for gas and motel expenses. Our decision to marry was firm. We got underway about 3:00 a.m., stopping at a restaurant a few hours later. Katrina's love encircled me, and a voice whispered of a secret destiny and a journey within my soul.

Although we'd been up all night, another previously silent part of me was awakened. My intuition was clear as I followed the direction of my inner voice. It was communicating with me in a way such that I could comprehend its higher meaning.

Arriving in Carson City, Nevada, we found a motel and stayed long enough to make love. I heard a strange symphony—it contained a song with a message to all of those who have trodden a peculiar path of love. It was a message about the inner trials and tests of the Christ consciousness that lay before me. It was about a future time—a destiny that unites mankind with a great hierarchy. This musical, mysterious thought was with me while making love. The message suggested there was something

I needed to do and complete. To me this meant to unite with an inner circle of cosmic individuals who were beyond our own world.

Katrina was a path, and through her I began to journey and fulfill a special destiny and certainly something unfinished. I perceived and interpreted an inner subconscious language. They were visual impressions of various geometrical forms and symbols. I was intuitively communicating with a language of live and vibrant symbols. They were thoughts of conscious forces.

Katrina and I were married in Carson City in a small chapel. I became flooded with the sensation of unity. The Christ light within us was a witness to our marriage. When Katrina gave me a ring with a red gem, emotion rose within me and I began to tear up, silently, unsuccessfully trying to conceal my feelings. She radiated with the soft glow of a full moon that lit the night sky around us that opened my heart to a memory of a great teaching. This teaching was where all things came alive within my mind and soul. As we left the chapel, the lady told us that the fee was twenty-five dollars. The thought of money rushed through me—suddenly, I was in search of something that was foreign to my higher self. Between us, we had the exact amount.

We left the chapel without any money and drove to Reno, where the wind was blowing fiercely. We stopped and walked through Harold's Club, and I felt at one with everyone in the building. All eyes seemed to be the eyes of love—the same love that radiated from Katrina. We were lost in the maze of people. They seemed to be like the stars that occupied a certain space of our universe and felt at one with that part of the universe.

Then—all at once—the obverse manifested itself. Where I had been far above, then, I found myself far below. It was the magical inner power that comes from above and is manifested within our hearts below. Whatever is done here has been completed, I thought. But my understanding was coming from above.

Suddenly, the casino was transformed and its worldliness became unbearable. We asked several people for the way out. A pathway was cleared and we were able to escape that worldly influence. The people in the room glittered in silver and gold. I did not want to spend any more time in this particular section of their starry universe. Passing through the doors, I felt a distinct relief of pressure. The turbulent west winds at our backs lifted us in the very spirit of love itself. I was filled with the idea of a supreme destiny.

We left Reno, heading again for my parents' home. I recall filling the car with gas at the Shell station in Lake Tahoe, California. It was early evening and we had not slept since we left Felton, yet I was wide awake. A thought came to my mind concerning the peculiar pathway of Christ's crucifixion and the day of world ascension. My vision turned into clear images that were entirely symbolic to me. These images conveyed a secret destiny that our lives were to follow.

While at the gas station, I noticed the giant yellow "Shell" sign. It was symbolic of the planet Venus to me. This image was an introduction into another world. Again, these things were symbols and a language of a peculiar knowledge that had forged itself from the fires of an ancient memory.

I was anxious to get back home and introduce my new wife to my parents. I wanted to share my revelation with them. We stopped to sleep alongside the road. I began to doze for a moment and gazed toward the night sky, then drifted into a deep sleep. I slipped out of my body, soaring through time and space, and was vacuumed into a small group of stars that shone above the quiet roadside. Waking, and not knowing how much time had passed, I started the car and drove off without disturbing Katrina.

Feeling oddly re-energized, I continued the five-hour drive to my parents' place. Upon arriving, we tried to step quietly, but my mother awoke immediately, followed by my father. They were anxious to meet my new bride. We spoke for a short time, and then went to bed. Except for the brief nap along the road, we had not slept since Thursday and it was now early Saturday morning.

I saw Wanda put down the file she was reading. She rolled her head and I began to massage her shoulders with both hands. "Dan, let's take a break. If I don't get outside and walk around a minute, I'm going to go stark raving bananas."

"Jesus, Wanda, I'm running out of time." I made circles in the air with my arms again. I paced for a few moments, sighed, and stared out the window. I hardly heard Wanda. I picked up my copy of the

Koran, which was dog-eared and filled with at least 150 annotated page-markers, and began reading.

"You should take this thing in small bites, Dan," Wanda admonished. "If you become consumed by it, we're lost. I'm afraid you'll check out of the planet or something."

"As a matter of fact, I was just thinking about the same thing. Going over this stuff again puts me right back in the center of the craziness and sometimes I wonder where it's all going to end."

"What were you going over in your mind?"

"Well, it's about that time I dreamed I met Ritter von X. I was actually speaking German with him and I don't know a word of German. I even saluted him as a Nazi, Wanda."

"And then you actually met him, years later, right?"

"Yes. Remember, I think I told you about the phone call."

"What phone call? I don't think you did tell me."

"It was when Diane and I were doing some research at the library and I found his name in the Encyclopedia of Associations. I just called him up and told him about my research on the theories of the Hollow Earth. That started the whole thing with him. I told you about this—he invited me to a meeting. We talked. I taped our whole weekend of conversations. Diane transcribed them. She ended up flipping out and she walked out. And now, here we are."

"And now here you are, you mean. I'm just a 'ghost' here, remember?"

"Maybe we're all just ghosts, Wanda—working for the higher purpose."

"Yeah, right. Well, I think I'll just float out of here for a while. Max will be home from work pretty soon." She headed for the door. "Where's it going to end, Danny?"

"It's not. This is only the beginning."

The phone rang as Wanda was leaving. I grabbed it as I watched her wind through the trees to her own property. "Oh yeah," I told my caller. "The book—sure, couldn't be better. My new next-door neighbor from Denver is helping me with it. We've just about got it nailed down. Now, what can I do for you?"

Chapter Six

❦ The Earthquake ❦

And behold, there was a great earthquake: For the angel of the Lord descended from heaven and came and rolled back the stone from the door, and sat upon it.

—St. Matthew, Chapter 28, v2

One Sunday morning, when Katrina and I were temporarily living at my parents' home, an exalted peace came upon me, along with an enormous influx of power. I was receptive to impressions from various levels of consciousness. These impressions manifested as intuitive directions for me throughout the day. I had always known that a woman's unconditional love would enlighten me.

However, as the day moved on, the profound peace that radiated inside me diffused, and I began to receive negative impressions from people all around the world. My mind seemed to be a receptacle, a psychic garbage dump soaking up these negative actions that would unknowingly result in certain disasters. Once seeded, these negative bombardments would sprout into patterns that would soon affect the surface of the Earth. These were evil ideas contrary to natural law. I became extremely alarmed at the thought that the Earth and its people was being tampered with by governments and self-proclaimed entities unwittingly destroying the delicate balance of cosmic forces within the mind of man.

Many negative subconscious seed thoughts began to germinate their destructive destiny. The planet Earth and its inhabitants were summoned by these negative controlling forces. My intuitive receptiveness and perception of those forces were focused, clear, and sharp as a spear. I was aware, conscious, and awake within the atmosphere of the Earth. I was of one mind within the consciousness of the air. Air is another form of matter. Air is a pure consciousness. These forces polluted this pure form of consciousness and it was my job to purify them by the light of the Creator. My awareness was like an angel of air purifying the gross and dense matter around me.

My understanding of this negative consciousness was that of a continuation of the age-old battle between the origin of creation and the fall of man from the Paradise Earth and then the destruction of Paradise Earth. I realized my responsibility to Earth, mankind, and the Creator.

I began to ascend to a higher degree of awareness that was magnified by the Earth's atmosphere. It acted like a huge magnifying glass and gave life to my intuition. This consciousness allowed me to direct these negative patterns. These suggestive forces bombarded my entire being. It was like meteors colliding with a planet's surface. Some were military in origin and others were individual. I was a mediator and a balancing factor between the planetary karma of the Earth and the will of the universal consciousness. I needed to reconnect with the source that decreed the image of my identity and destiny. I was moving toward my ascension through the veiled and invisible forms of matter. My soul was climbing through the fixed veil of the starry universe. I could now see my soul's path beyond time and space.

Time would slip and sometimes stop altogether. In my mind, space and time would fold away. It was never really clear whether now was now, or last year, or ten years hence. Time was only relative and space never existed.

That happened to me one Sunday morning in 1990, while I listened to Dr. Stranges' voice—recorded in June of 1989—over the small cassette tape recorder at my dining room table. I did not know, do not know, who Dr. Frank Stranges is. Nor do I care. I have not read either of his books about UFOs, flying saucers, and his alleged friend from Venus, Valiant Thor. Yet, early in 1988, I would be drawn to a small blue book called *My Friend from Beyond Earth* by Dr. Stranges. And when that time came, it would be because of how I was feeling then, in 1990. It would be a direct result of my need to link man with the source that decrees the image of his true identity and destiny. My destiny is the journey where soul returns to the origin of the Creator.

Sometimes, things just happen that way and this was one of those times for me. I am transported back to 1989, June 25th to be precise, listening to a tape. Dr. Stranges's voice was coming through loud and clear:

> Bear in mind again, that before the great flood in the days of Noah there was a water canopy that used to exist over this planet—before it fell down and formed the oceans as a flood. The Bible tells us plainly that there was a great ring of water or an ice canopy all around the atmosphere. We read in Genesis 1:7, 'God made the elements, the air or atmosphere, and divided the waters of the Earth from the firmament Heaven and God called the dry land, "Earth." Job tells us that the sky was a beautiful blue at all times. No storms, no cyclones, no tidal waves, nor even any clouds for 1,650 years. It never rained! A mist went up and watered the face of the ground in Genesis 2:6. The great ice canopy melted and fell and the first rainbow appeared in Genesis 9:13. Before the flood, the light of the sun was as even and as perfect as God made it and the climate was cool and the temperature was the same all over the planet...

The day settled into the evening as the sun slowly sank and disappeared below the horizon. A great magnification of universal consciousness was enveloping me. I was in a state of oneness and all things expressed the image of man. Upon my forehead was an opening into the inner workings of the universe. The internal workings of the solar system were being received by the magnificence of the "third eye." Within this conscious sight, I could perceive many things behind the actual human records of the past, present, and future. This I understood to be a glimpse and meaning of the eternal, the will of the Creator's light. In part, I was in a position to direct these cosmic forces.

This was another great understanding of the inner workings of the Creator. I felt at one with the absolute law that governs these workings. This absolute law governs and sets the universe into motion and into eternal harmony. I believe that I have been privileged to have had a glimpse into the Creator's illumination.

Many earthquakes have been predicted by modern prophets of our century. Some were about to take place again, as a method of growth and to awaken human consciousness. I recalled the conflict of the light and dark aspects of the Atlantian continent. The memory was cyclic in nature. Its origin was a tremendous seed thought from an ancient time. The memory began to seep into my subconscious mind. The subconscious mind is universal in all people. I recognized this ancient memory and knew it had to be altered and changed in the minds and hearts of every individual on Earth. If this memory was allowed to surface, mankind would meet Earth's plane with an unfathomable force of destruction. The memory would rip the Earth and send its surface existence careening under the sea.

This cosmic cycle needed to be neutralized and balanced within human consciousness. The creator's light resides within those who have their souls grounded in the heart of the Earth and the divine love that emanates within them.

This ancient memory cast its great shadow of destruction upon the Earth. It was like a visible and predictable comet. Only those who do not bend or yield to the treacherous karmic winds of the past could neutralize this Atlantian memory cycle and its destructive power. Individuals like me had to remain awake and aware of the cycle's mighty power over mankind, without being swept up into earthly chaos by its swift current. This ancient memory has been coercing us for centuries into repetitive and destructive cycles in order to perpetuate itself and prolong war, disease, and the continuous cycle of death. It was time to overcome these cosmic forces in our time.

I did not stand alone. Many cosmic beings stood in unison, as a temple of great light. We were opened to its greatness and receptive to its will. We had been raised up into the Creator's light and given authority to direct the Creator's conscious will through us. We stood in the name of the deity that created the heavens, the Earth, and all worlds to come. We withstood the ancient karmic force that carried the subconscious seed memory in a cycle of many incarnations. But the winds of the subconscious past did not destroy the surface of the Earth because the world karma that surfaced in memory became neutralized by divine love and by the handful of these rare individuals. It was clearly understood that what human reason and intellect could not accomplish, the cosmic could.

Just as suddenly, my consciousness was left swirling, crashing through the memory of recorded time. Descending back into earthly senses was not so easy. In an instant, I lost all insight of what I had just been given. I walked into the bedroom like a child who had been unfairly disciplined. I began to sob uncontrollably. Katrina, following me, asked what was wrong. In a tearful state and feeling as though God had abandoned me, I told her that I was falling from my higher self, entering into the world of the senses. I explained that I was falling back into an earthly state of awareness—awareness based on human senses alone.

Emotionally distraught, I went into another room and closed the door behind me. I sat alone upon the bed. Without Katrina's knowledge, I reached for a pistol I had hidden in the room and pointed it at my head. Mentally, I walked into and through yesterday's nightmare of fear and shadow, entering a haunting memory shrouded by the wrath and fierce will of Nazi Germany.

I managed to pull the pistol away from yesterday's shadow and away from my head. I pointed the 9 mm P38 out toward the open window of the bedroom and fired it once. Many secrets of my hidden self were revealed. Something had to be done, and fast.

Time slipped again and I found myself in the future. This time it was 1983. It was 3 a.m. and I was on the phone, listening to the thick accent of Ritter von X:

> It also destroyed the thick layer of water vapor that was in the atmosphere of the Earth. While it was in existence, this canopy produced what scientists called a greenhouse effect upon our planet Earth. The water, its vapor insulated, provided a uniformly warm climate all around the world.
>
> Also, this vapor canopy shielded the Earth from the sun's most harmful rays. Vegetation was wonderfully lush and healthy. And plants were so nutrient-laden that they supplied all of human nourishment. Of course, this made humans vegetarians, prior to this great catastrophe.
>
> Today very harmful rays are received from the sun. Radiation causes genetic mutations, and it also plays a significant role in the aging process. Yes, we grow old because of the sun. In the center of the Earth, there is a type of sun that does not cause an aging process. It is most beneficial to the balance and longevity of human life. However, when the great catastrophe

occurred, one of the two moons surrounding the Earth crashed into the surface of the world, sending it careening and spinning. All things were changed. Gravitational force suddenly changed. Great cold and heat were applied in sporadic patches—a cataclysm that can scarcely be described.

Just as suddenly, I was home with Katrina, who was completely unable to comprehend me by now.

The path of consciousness and the human spirit trod ahead and mirrored this memory before me. I began to feel angry at the human race; at the same time, I needed attention from Katrina. But before I could indicate this need, Katrina received a phone call from her old boyfriend, Peter. His message was for me to meet him in a nearby ghost town early that evening. In a panic, Katrina told my mother about Peter wanting me to bring my gun and prepare for a duel. Once again, I was greeted with another outrageous karmic pattern that had to be overcome. Always, the greater consciousness was testing me.

In a rage, my mother forbade me to do something as stupid as having a gunfight. She called my brother, Fred, and he came over immediately. I met Fred outside in the driveway. Before he got out of his car, he slammed a magazine into his pistol, saying, "You stay out of this battle. This one is on me." He was like a knight and a servant—willing to do whatever was necessary for my protection.

I went back into the house, trying without success to comfort everyone. I knew I had been time-walking. I knew I had passed through many karmic events. I was on a journey back to home consciousness. Blindly, I walked back to the bedroom, laying my hands upon my face. I was deeply saddened, realizing that rebirth of the higher self could take hundreds—or even thousands of years.

I cried in a fearful voice, knowing that I could not remain in a state of cosmic beauty and eternal power and bliss. I reasoned that this, too, was the will of the Creator. Katrina tried to comfort me without really understanding what was happening.

As night fell, I could feel the coldness and the absence of the warming sun. Life was exiting by day while death seemed to creep in by night. My ideas of time drifted as I fell through a long, twisting tunnel of darkness. The night sky swirled as I began to journey and fall through darkness into an unknown day.

I had a dream about an enormous earthquake. Everything in my mind was shaking. I was falling to the center of the Earth. When I awoke in the morning, an earthquake was actually taking place. I felt my consciousness shift and slide, followed by a struggle for stillness. A new and deep crevice had been opened and my subconscious released another mystery of my being. I awoke in fear of that memory and that it would surface in a magnitude not recorded on Earth before. My heart feared this the most. It was vital that I live in accordance with the decree of the laws of light so that my family would find absolute refuge and safety. I was now slowly awakening into my earthly senses but with a conscious memory of the earthquake's destruction. My fear lingered like broken cosmic particles in a wounded space and time warp. Once again, I descended back to Earth and memory, shocked from my cosmic fall. Before I could recover from this awesome cosmic experience, I was triggered by yet another memory. It was Sir Isaac Newton's third law of a cosmic aftershock—for every action there is an opposite and equal reaction.

My parents were preparing to leave for dinner at some friends' home. Oddly, my mother handed me some clothespins and asked me to put them away. When I felt the contact of the clothespins in my hand, I received an intuitive image of a large earthquake. I controlled my inner panic. I was silent as my parents left, and Katrina and I left a little after they did. To this day, when I pick up clothespins, I am instantly haunted by the memory of a major earthquake.

Throughout that day I experienced opposing states of consciousness. The first was how to process, understand, and clarify the deeper meaning of prophesy, and how to alter certain destructive cosmic subconscious forces that reigned against divine will. Perhaps the most-respected and best-known pure psychic of our time, Edgar Cayce, known as the "Sleeping Prophet," predicted unusual cataclysmic events, including cataclysmic earthquakes of unfathomable proportions that would send most of the western United States under the sea. New Age people in great numbers believed in his prophesy and have prepared for this coming destruction.

These cosmic hosts and I were united as one entwining eternal flame and rose above this subconscious cosmic ancient memory and neutralized its mighty forces of destruction. It was by divine grace, that I was privileged to participate in this cosmic event. It was by the action of these cosmic hosts that Edgar Cayce's ancient destructive memory remains eternally asleep, never to awake and reveal the subconscious horror of the memory of Atlantis in our time. There are many cosmic subconscious memories best left within the cosmic garbage dump. Man has a grave tendency to try to raise his past consciousness at a cost of destruction beyond the stretch of human imagination. This is the price mankind must pay before delving into the evolution of his consciousness. To the New Age movement I propose a vision of the Golden Age rather than repeating the greatness of a fallen memory. It is best to leave the angels of destruction alone—but never forget them. Remember, "Those who do not remember the past are condemned to repeat it." (George Santayana, from The Life of Reason, 1906)

We drove to a nearby shopping center, where a store was having its grand opening. It was packed with people. I wanted to rebuild my inner self, and found myself being led intuitively to a particular painting. I walked around until I found the very painting that my inner self had sought. Removing it from the wall, I exclaimed, "I found it!" I bought the painting, a three-by four-foot piece of art with a large, descending triangle at its center.

The symbol of the triangle released the very core of my being. A magnificent teaching was being communicated to me by means of this triangle. The triangle became an endless storehouse of knowledge that radiated from its magnetic center. Leaving with the picture under my arm, I felt a sublime peace. It infused my entire body. My memory haunted me again with mighty earthquakes that were to shake the surface of the Earth. Again, the earthly karmic cycle of the Atlantian memory began to rise and take form.

Remaining silent about this inner activity, I left the store with Katrina. We stopped at a friend's house but no one was home. I felt like telling people about the earthquake, but something prevented me. We drove through my hometown. We were like a newly created man and woman. But an angry karmic cloud remained over us seeking destruction and compensation.

I felt the breath of God's annihilation upon me for the wrong that humankind was doing in His garden of creation. I began to fear His presence. At the same time, I began to see His inner kingdom as a golden setting sun. No shadow of any kind dwelled near the Creator's radiant presence. I traveled farther. The God of creation was with me. But where do I go? I thought.

While we were on the freeway, a car passed us as though we were standing still. The driver's face reflected a deep fear and panic. He seemed to mirror the beginning of a cataclysm. The buildings and movements of all things on the Earth began to vibrate in terrible disarray. A tremendous consciousness began to intensify and encircle the Earth like a serpent squeezing its victim. The brake light began to flash on my Volkswagen's dashboard, indicating that I had lost all braking power. Fortunately, I was going uphill toward San Jose, California, on Highway 17. I pushed my brakes to the floor and shouted to Katrina, "We have no brakes, absolutely no brakes at all!"

Katrina sank into a deep silent panic. She clung to me while hanging onto a thread of faith for survival.

My thoughts were immersed in an unknown world. Intuitively, I began to overcome this snakelike consciousness that encircled the Earth. Our human consciousness is the very fabric of the vast starry galaxy of the Milky Way as far as the eye can see.

When I spoke, my thoughts transformed into the material by the power of an unknown world. Suddenly, my brakes returned to normal. As I rose higher into the realms of cosmic consciousness, a fear of earthquakes slowly consumed me. All at once, my thoughts and attention slammed back into reality.

Then I saw my brother, Fred; pass me on the other side of the freeway. My fear of earthquakes began to intensify. His subconscious mind mirrored the Atlantian memory. Again my brakes failed. The road was curvy and dangerous and there was no safe place to pull off and stop. I was going downhill now, staying with the flow of traffic. In an instant, gravity would disappear with a warp in space and time.

My fear began to take control. I was in a state of panic, but remained silent. Finally, I asked Katrina, "Doesn't Fred know that the continent is about to be entirely devastated by mighty earthquakes?" Instantly, my fear manifested on a physical level: my brakes were completely gone and the highway remained downhill for several miles.

Unaffected, Katrina softly replied, "Don't you want peace in the world that you love and live in right now?" Katrina's faith in peace became my battle with its adversary, the "god of this world."

The Bible refers to Satan as the "god of this world." However, it is not saying that he has ultimate authority. It is Satan that rules over the unbelieving world. It is in 2 Corinthians 4:4, the unbeliever follows Satan's agenda. It is according to that passage, the "god of this world has blinded the minds of unbelievers, so that they cannot see the light of the gospel of the glory of Christ." Satan's agenda includes pushing a false philosophy onto the unbelieving world—a false philosophy that blinds men from the truth. Satan's philosophies are the fortresses in which people are imprisoned, needing to be set free and brought captive to Christ in obedience to the truth.

I thought, "How could we leave our families with a fear of a mighty cataclysm?" The sound of her gentle, sweet, true voice began to swallow the fear in my mind. The car began to slow as we approached level ground. I was at one with the great universal cosmic mind.

I began to slow the car with my emergency brake and drifted to a stop alongside the freeway. The journey to another world was temporarily altered. I was being prepared to battle the world's adversary of peace—the serpentine deity at the heart of the Earth. I realized a new challenge was upon me. It would take a meeting with God—face-to-face, to free the soul of humanity from the bonds of this treacherous, ancient, and worldly deity.

Evening drew near. I drove to a nearby Shell gas station. Again the shell represented a rest in consciousness. The memory of the Golden Age flashed before me, intact. A blinding white light encircled my head. I slumped in the driver's seat exhausted but at peace for the moment. Yet a vague awareness in the pit of my stomach nagged. This moment of freedom from Katrina's loving words would give way to yet another adversary.

Once again, I traveled forward in time. I seemed to be whirling and spiraling into the upper atmosphere. Below me was the Earth. A white, spiraling light surrounded me, lifting me upward, upward into a vast universe. Traveling in time can be an incredible ride. It is probably like being outside of your body, yet being aware of everything around you. Traveling in time feels like being an astronaut without a spaceship. I could see the Earth below me like a beautiful blue and white ball with patches of green. Then a white, spiraling light suddenly slammed me back onto the surface of the Earth, carrying me to a new dimension. I was now in that new dimension, talking on the phone to Ritter von X for the first time. It was a sweltering Indian summer night in 1983, and the sound of hundreds of crickets filled each pause in the conversation:

"Excuse me—ah—well—I located your name and address through some research I have been doing in the library." I was tentative, unsure of what his reaction would be.

"Yes," responded the raspy, guttural German accent on the other end. "I have your letter in hand."

"Yes?"

"Yes. That is what I have just said." Ritter von X was to the point. Only the crickets responded. "What do you have in mind?" he finally asked.

"Well, I was really surprised to find this organization in the Encyclopedia of Associations," I replied, excited to actually have him on the line.

Ritter von X (a nom de plume at his request) 1943-1945. Involved in the Reich Undersea Boat service aboard U-530. In 1969, he recovered the Sacred Lance (Spear of Christ from Antarctica). In 1979, he helped plan the Hartmann Expedition to the Antarctica to recover Hitler's secret treasure and invested as a Knight of the Holy Order of Knights and of the Sacred Lance.

"Yes," replied the German. "We are an international organization, doing very important research."

At last I got control of myself. "I've been doing personal research, studying the ancient Icelandic sages. I visited Iceland last year, and lived there for the summer."

"Oh, you know about Thule?"

"Yes. I've been reading things about Thule in Gardner's book, Moon Gate."

"Ah, yes."

"And I read The Smokey God, as well."

"That was an astonishing work, The Smokey God. Based on fact, I think," von X mused.

"This is why I'm talking to you—why I called you. You are my first real live contact other than the research that I've been doing for some time. I just have some questions that—"

"Well, I hope I can answer them, surely."

"The first question is, I'm just wondering how close you are to your objective—an expedition."

"We sent an expedition in 1979 to the former territory of Deutschland, in Antarctica. It was successful. In fact, we have written a manuscript about the mission." There was a touch of pride in von X's voice.

"The Smokey God mentions the people of the interior. I was wondering what additional information you have concerning them."

"Well, you know—" he paused and laughed. "The subject is little known except by certain groups in the world, and, of course, the American, Admiral Byrd. He actually entered the realms by accident. He attempted to keep secret what he encountered there. All of those records have now disappeared. And I think it was in 1947 that the American government sent a rather large expedition. Some ships and aircraft were utterly destroyed. They ran into a force they could not comprehend. This was kept highly secret. Of course, the Russians currently are very interested in watching this area of Antarctica as well as the English and Americans. And there are—how do you say?—super races, and there are flying saucers that come and go from certain entrance points there. This is a well-known fact. With every attempt to intercept them, the people faced a terrible disaster. I know you may think this is fantastic. The terrible weather now being experienced by the United States in particular is because of the government's tampering and trying to attempt to enter this forbidden area. Do you believe this?"

What could I say? "Well, what I can support in that belief is my own personal cosmic experiences. For some years now, I've been led out of the darkness, so to speak, into a greater realization. I know it's intuitively leading me in this direction."

"Yes, they are there. They have been there for a very long time. They are not interested too much in the affairs of the world out here because they are so far advanced, and they regard ordinary humans as barbarians. In all our research we have learned things that boggle the mind. There are agents in the world who do not want the people in the center of the Earth to know that they are not alone, on or in it." He laughed again. "The U-boat 209 was sent on a special mission to follow a certain course that was highly classified. I have a letter that I received from one of my old friends who was on the U-boat 209; in 1943, I believe it was—eight years before the letter arrived. I didn't take note of it immediately. But when I opened it, to my surprise it was from my old friend, who stated that the 209 had entered into another realm, and they did not transmit and they could not return. I will send you a copy of that."

"I will appreciate any information you could update me with."

"You have got some other questions, have you not? This call must be costing you a lot of money."

"It's well worth the money," I reassured him. "I'd say you are my only live contact on this."

"Well, there's a society existing in the world today to which many hundreds belong, Americans and Europeans and even some—and even the Russians are interested. We know the Russians are aware but we have no contact with them. The society's members, however, are scientific people."

"One of the other questions I have is how could world peace ever be realized?"

"Well, we had this idea in the group at the time it was formed that we would devote ourselves to an American modern-day search for these people. What we have in mind is somehow, someway to contact them in a very positive manner and have them help the surface civilization to come to an understanding that this is the solution to humankind's problems. In fact, this scientific progress is essential for understanding. We hope someday to perhaps send a delegation before the United Nations—an unprecedented action—to say that we are from the interior of the Earth, but we are here to cooperate. But you see, they have remained alone and closed to us all these years until recently, when experiments with atomic bombs began to affect them. Perhaps they are in the air now. We believe that a nuclear war might break out, but we think that they will not let it occur. They have weapons, machines, and so on. They are amazing, really. In comparison, we are barbarians. Do you understand?"

"That's understandable. In The Smokey God, Olaf Jensen left some maps."

"Yes, that was in California, around Los Angeles."

I paused to think. "From what I understand, he left these things with the author, Gardner, who mentioned that he was going to leave them to the Smithsonian. I called the Smithsonian and there wasn't anything there."

"I wouldn't expect you to find it," said von X. "I have to speak frankly with you. There will be no records of these things because the Central Intelligence Agency, the KGB, and others do not want this information to be available due to the very nature of the organization."

"Now, these would be the people who would be suppressing the truth, would you say?"

"Oh yes, very positively and most definite. But it is classified information and unavailable. As you know, Admiral Byrd mentions that he entered the opening and observed animals in there. They took photos. The photos have all disappeared and of course, Byrd, you know, was made to appear as a madman. They discredited him totally; he was a victim of a very powerful group of people that I cannot mention it this time."

"I have information that in 1947 he was in the North Pole," I said. "But I cannot find any information concerning his flight to the North Pole. I went to every major university and I can't find out anything about this."

"All of this information has been withdrawn little by little, without people even realizing the significance. That's what I'm saying to you about this information. No one will ever know that this is true."

"How did you first get to this information? What was your beginning?"

"It was from the U-boat 209. I'll send you copies of the letter that will explain many things."

"Is there anyone you know who is the leading authority on this subject? Or perhaps that person is you?"

"Oh, no, I'm not a leading authority. But I belong to an organization in Germany that has been doing research on this for a long time."

"Are there any study groups doing research that's a little more detailed?"

"Yes, there are, but it's all closed."

"Are there any members here in California?"

"Yes, there are several. Unfortunately we can't disclose their names because some of our researchers have been heavily interfered with and tampered with, and some of our people have been harassed. Frankly, what you are interested in is something that is so strangely, beautifully true, yet it's very dangerous territory. If we are not extremely discreet about it, serious consequences would occur."

"Yes, I agree. I, too, feel it necessary to keep things very tight and closed. As I say, it's been very personal research for me. And I've shared this with only a couple of people who have supported me in the research as well."

"Well, I can tell you one thing, Mr. Weiss. These beings are so advanced that they have little to do with the affairs of the surface world. But now, atomic explosions have interrupted and disturbed them. They have changed—they are in danger. They are really not intellectual geniuses—not that they don't exist. Some people here are from there. They are all blonde and blue-eyed people and much taller."

"Now, The Smokey God mentions a certain stature, a certain size. Are they the same size?"

"The book said they are of enormous height. Actually, they are taller than average in size. I would say they are about two meters. They are very tall blondes."

"My reason for pursuing this is not only to continue the search and prove the truth of the information I'm receiving, but mostly because there is an interior, an intuitive motive that's somehow pushing me along. And also because of certain different psychic experiences I've had that are directly related to this theory."

"Yes. I understand, absolutely. But there are things that I cannot and will not transmit to you by phone or even by letter but only person-to-person. Then I will tell you many things—things I am working on and other things that are happening. Unfortunately, I dare not put this information into a letter. You have a good understanding and grasp of what is happening."

"I am definitely looking forward to learning from you what literature you have, and I'm definitely interested in meeting you."

"That's fine. Perhaps we can get together soon. This afternoon I will make a point to obtain a copy of the U-boat 209 expedition letter and send it to you."

"Do you know when would be a good time to meet in the future? I can take a plane."

"I will be gone in October for two weeks. Perhaps we could arrange something around mid-November."

"Okay. I will look forward hearing from you by mail, and if you have time I would like to meet you in mid-November."

"Very well, then, Mr. Weiss. I will write you and let you know when we have the time available."

"That's excellent, sir. Thank you."

"You should prepare an itinerary. Meanwhile, it's been very much a pleasure to talk with you."

"Yes. Thank you very much. Goodbye."

Ritter von X and I maintained regular communication after that. I would, in fact, meet Ritter von X at the prescribed time and tape an entire weekend of in-depth conversations with him. Years later, I was to remember three predictions that came to pass. The first was in a letter dated January 22, 1989, with the words, "Also, please take sufficient precautions there. You may have a major earthquake within the next three months or so. Thal states that there is a major sodium hydro collision coming beneath the California crust, so please be advised." The earthquake indeed took place; just as Ritter von X had said it would. Thal is from the inner earth and commander of the Flugelrad (UFO) squadron in the Northern Hemisphere. He visits the surface world on certain occasions.

Later, on October 30, 1989, I received another letter stating, "The year beginning with 1990 WILL be filled with awe and with the developments of certain factions in Europe. Interesting, too, is the fact that Gorbachev has secretly converted to Christian beliefs, and will call on the Pope early this November for instructions…thus you have the force behind the so-called *glasnost*, or openness."

On November 14th of that same year, another letter predicted the fall of the Berlin Wall. "I watched with fascination on TV as *der Mauer* (the Wall) was attacked and broken by jubilant German throngs, not east or west, *but Germans*! We spoke on the phone shortly before this occurred on the 9th. Indeed, he had spoken with me of these things in a previous phone conversation. The mystical bond we shared would not be taken lightly by either of us.

Chapter Seven

❦ CONVERSATIONS WITH ❦
RITTER VON X

DAY 1

Verily I say unto thee, today shalt thou be with me in paradise.
—Jesus Christ, the Bible, St. Luke—Chapter 23, v. 43

IT WAS THE second week of October, 1983. I had returned from a fascinating weekend meeting with Ritter von X—finally meeting him in person. Diane was still helping me with research, so I had asked her to transcribe all the tapes of my conversation with the knowledgeable German. She was at the computer, ready to begin, when the letter arrived.

"Diane. I got it!"

"Got what?"

"It's the letter from Ritter Von X—about U-boat 209."

"My God! Is it in German?"

"Yes, with a translation on the back."

We read the letter together:

> Dear old comrade! This news will be a surprise for you. The submarine 209 made it! The Earth is hollow! Dr. Haushofer and Rudolph Hess were right. The whole crew is well, but they cannot come back. But we are not prisoners. I am sure this letter will reach you. It is the last connection with the submarine 209. We will meet again, comrade. I am worried for everybody who has to spend his life on the surface of the Earth since the Fuhrer [Hitler] is gone. God bless always our Germany with hearty greetings. K.U

Herrn Willi Schaus
203 W. Love ST.
Mexico / Missouri

20. April 1947.

Lieber alter Kamerad,

diese Nachricht wird eine Überraschung für Dich sein. Das U-Boot 209 hat es geschafft. Die Erde ist HOHL! Dr Haushofer und Hess hatten recht. Der gesamten Mannschaft geht es gut, aber sie kann nicht zurück kehren. - Wir sind kine Gefangenen. Ich bin sicher, dass diese Nachricht Dich erreichen wird, es ist die letzte Verbindung mit dem U 209. Wir werden uns wieder begegnen, Kamerad. Ich bin in Sorge für jene, welche ihr Leben auf der Oberfläche der Erde zu verbringen haben, seit der Führer gegangen ist.

Gott segne immer unser Deutschland.

Mit herzlichen Grüssen,

Karl (K.C.)

Letter in German to Ritter von X from K.C. about U-Boat 209

"Whoa! So it could all really be true?" Diane sounded awestruck, and somewhat frightened.

"I'm not making any judgments here about truth just yet, Diane. Let's just look at this research objectively. Otherwise, we'll get caught up in the same age-old conflicts over truth and mendacity, lightness and dark, good and evil. That's what we're trying to get a perspective on, remember?"

I tried to be the voice of reason—instructive and calm. "One thing I know is true: more than one person believes this to be true. Remember when David was doing his master's thesis on the Nuremberg trials?"

"Yes, that's right. He's the one who dug up that book by Harbinson, isn't he?" Diane asked.

"Yes. It was called Genesis," I remembered. "In his author's note, Harbinson talked about what was thought to be a secret weapon the Germans had developed called Foo Fighters by the British—mysterious little silver balls racing alongside the wings of fighters flying intruder missions over Germany. Pilots encountering them reported that they seemed to be radio-controlled. They mysteriously disappeared right after the war. Some people think they were UFOs.

"But other people think they were a secret German weapon. The Air Force's Project Blue Book speculated on that issue. In 1956, Captain Edward J. Ruppelt wrote that after the war, we discovered that the Germans had several radical types of aircraft and guided missiles under development. They were the only known craft that could even approach the performances of the objects reported to be UFOs. During the war, Hitler wanted to get into Antarctica, so he sent a Captain Alfred Richter to the coast due south of South Africa. During their daily flights, they discovered vast regions of the South Pole surprisingly free of ice. That's the same thing the Byrd expedition found in 1947. I've got that Byrd article in the National Geographic in here somewhere."

Diane was anxious to get on with the monumental job of transcription ahead of her. "Later, Dan, I've got enough to think about right now with the Ritter von X tapes."

"But he was on one of those boats! He was on the U-530 that was launched from a port on the Baltic Sea. They had members of the flying saucer research team with them, along with notes and drawings for the saucers and the designs for the gigantic underground complexes and living accommodations in the Harz Mountains."

"Where did you get all this information?" Diane asked.

"I got it from Harbinson's book. But later, when I talked to Ritter von X, he confirmed it. The book also says that the U-boats surfaced mysteriously off the coast of Argentina, two months after the war was over. The crews were interrogated and sent to the US. Two years later, the Byrd expedition was launched. 'Operation High Jump,' they called it."

"Interesting—there seems to be a strange puzzle here. How does it all fit together with Byrd, Germany's interest in flying saucers, underground complexes, discoveries of vast ice-free areas, Ritter von X and what do you make of it, Dan?" Diane asked, curiously, now less anxious to begin her typing.

"That's what I'm trying to figure out, Diane. Better start transcribing those tapes right away."

I went to the piles of books and file folders on the floor beside the fireplace and began to fumble through them. Finally locating what I wanted—Genesis by W.A. Harbinson, I flipped to a dog-eared page and began to read aloud to myself:

> "What was being suggested, then, is that throughout the course of the Second World War, the Germans were sending ships and planes to the Antarctic with equipment for massive underground complexes; that at the end of the war the flying saucer project's team of scientists were taken from Germany by submarines U-530 and U-977; that the Americans interrogated the crews of those submarines when they docked at what they had thought was a friendly Argentina; that the Americans, upon hearing of the Antarctic base, organized a military task force disguised as an exploratory expedition; that expedition was subsequently put into disarray when it came up against the extraordinary German saucers; and that the United States then pulled out of the Antarctic temporarily, in order to build their own saucers based on the designs found in Germany after the war."

Meanwhile, Diane began typing. Following is the major portion of the conversations, which actually took place during my weekend with Ritter von X on October 5th and 6th, 1983. Names and details, given to me in confidence, have been omitted. Because English was not Ritter von X's native language, some areas of this direct transcription are unclear. Ellipses and square brackets will replace these sections in the text.

> X: The air force was over Germany[...] that was the first scene then in those days of the Flugelrads in the air. It's an official German record of when the British and Americans flew the B-17, the flying fortress. These were the planes used at that time for bombing. Do you know that the Flugelrads fly up through those high-altitude formations, those tie formations, under them, over, among them, and pass downward at terrific speeds? And the Luftwaffe saw them also. They reported to the German high command and the Luftwaffe Center in Munich, almost at the same time, simultaneously. Germany announced that if this were a new type of weapon, then we would find the means to countermand them. Almost at the same time, the British had seen these also, and the BBC said in almost the same words—if this is a new German secret weapon, we will find out how to countermand it and destroy it.
>
> But, you see, these discs were traveling at such fantastic speeds that they were in the gun sights of the machine guns and so forth for a split second, and they were gone, so fast. Again, in the distance, they could be seen and observed quite easily. So the official records both from Germany and from England [show that] they were seen in those times. The ships were observing the war, bombing, and watching everything that had taken place, which they had done for a long time. This is record.
>
> The Americans had a name. They called them Foo Fighters. Did you ever hear that word, "Foo Fighters?" That's what these were. They were magnetic flying devices of such a fantastic nature that both the Allies and also what were called the Axis Powers of Germany—they saw them. And they had no answer for them. But Dr. Haushofer in Munich, Rudolf Hess, Heinrich Himmler, Adolph Hitler, all of these men were aware of what was taking place. And so you would hope and dream in some way to send the U-boat 209 on its mission to get the war stopped and bring order to the world.

This is difficult to comprehend, but they actually, finally, had hoped [confidential name] from the SS had even filmed these [Foo Fighters]. The only way they could film was to fire upon them so that the object could be seen. They had films on these. Then the Allies overran Germany in the end days of the war. It was a top mission of [confidential name] to get to the military complex near Spandau and to get those films. They had orders to destroy anyone or anything. The films were of the highest priority. Even the Russians wanted to know [the content of the films].

Dan: The Russians?

X: The Russians. They confiscated this material along with some of their top scientists. Secret negotiations took place between the English and the Americans, to negotiate under some commission that they had to create [an agreement] at Potsdam that all secret information was to be shared equally. The Russians did not share that information and in fact they thought it was part of Hitler's work in Austria where he was working on a type of magnetic flying disc. And this was the supreme mission of the Soviet forces: to capture the armada as quickly as possible so that they could grasp this, they thought. They thought that such a secret weapon was almost completely ready for use and if only they could capture that information, they would master the world.

Those films have never been recovered from anyone. But the Luftwaffe saw them, filmed them. The Americans, the English bombers somehow were foolish. They didn't film them. All they had was the testimony of the bombardiers, the pilots, and crew members on the bombers, saying, yes, we saw them. What were they? They went by so quick [...] they were round and they were shiny, and they flew through the formations with fantastic speed and maneuverability that defied aerodynamic credibility. They called them Foo Fighters. They couldn't believe what they were seeing.

Dan: In this book here called Genesis, he talks about that very same thing—Foo Fighters. He also speaks about the U-boats that were headed toward the Antarctica and they are saying that [...]

X: The U-boat 530 and the 977. You see, I was on the 530.

Dan: Oh, that's it right here, U-boat 530.

X: Look. Look here, my old souvenir. [Ritter von X stands and shows me his brass belt buckle that reads U-530.]

Dan: You definitely are a veteran of many interesting things.

X: You have all these cases where men are saying—and who believes them because they are looked upon maybe as lunatics or something?—but all of these reports of having contact with these people, these beings, cannot be just brushed aside as fantasy, because they actually have landed here, many times in various places, and conversed with people.

[With many] of these testimonies, I laugh because when I read a newspaper that says something like, it was some strange creature with claw hooks or something of this nature, and all of these things, you know what I mean, and they are not like that. They are tall.

They also have among those in the dwelling in the inner world, smaller people of various kinds. But these as I understand are a type of magnetic configuration. They are so advanced that they can even create beings to serve. Do you understand? They can create these beings to serve, as a race of workers, if you like, to do as they are told. They are like a computer. They are programmed to do this work. It is definitely a system, not of masters and lower classes and so forth, but it is like you have created a race of robots to do work, like in an auto factory. They have machines, robots now, to assemble most of the vehicles with not many hands even touching them because it's done with robot machines. These are inorganic machines, but it is our understanding that this race is called Arrianni. They have created these beings whose only purpose is to work and serve. They don't have any emotional feelings and [have] an intellect that is calculated [only] for specific tasks.

Some of the little people that we have spoken of have had contact with these Flugelrads. They are a type of crew and they are assigned a mission. They carry it out and they return with whatever is required. What is required of them is oftentimes a type of fantasy. You hear of people being abducted, taken away, and all sorts of strange experiments performed on their bodies, and all of these things, you know. Then the don't know why.

Dan: Yes. I've encountered them a few times myself. So these are manned spacecrafts that actually—? There is intelligence behind the intelligence operating the vehicles?

X: You have it correct. The intelligence is behind the intelligence. That's right. As I say, we understand they are emotionless. They have no feelings. They [have], perhaps, the faces of a canine and have a very high intellect.

Dan: This definitely keeps human understanding in check.

X: They have no desire, as I understand, to do anyone any harm, to hurt no one, but they are technologically defensive. In other words, they will not allow the craft itself to be boarded or taken over by force or anything of this nature. They only will use perhaps a magnetic energy to stop someone—causing them to black out—and then proceed with their mission without harm.

Dan: This is what I encountered just before I received a letter from you. I remembered a dream state about this very moment. I asked you several questions, and you are much the way I saw you in the dream state too. I remember asking you two questions. This is before I received your letter. I remember asking if you have been to the inner realms. The dream changed and I ended up talking to an unknown person, or individual being, or whatever was speaking to me telepathically. I was somewhere in Iceland at that time in the dream state.

X: There are entrances in the Nordic area both south and the north and also Dr. Haushofer's papers that strongly suggest that in Tibet were a great underground palace. Underneath certain palaces [are] some entrances. One of the entrances there, the Tibetans believe that certain priests can go and meet with the master of the world, as he is called. No one can really understand this mystery except in an abstract way. It's—what's the expression?—a boggling of the mind. When you realize that this is real, it's a place that really does exist. How can this be? We have scientific proof that the world is a solid mass with a molten core, and all of this. And then when certain forces are confronted with the truth, it revises your whole thoughts and ideas of technology, of what we are here. You no longer [...]. That's what you really begin to know when you have learned the truth. That's the old saying, "the truth will make you free." I never realized how much that was really one of the most philosophic gems of wisdom that has ever been stated. The truth will make you free. It does. It doesn't give you a sense of power, of smugness, like saying, "I know what you don't know," or something that way. It has a humbling effect. You really suddenly realize that there is more to all of this than has ever been thought. We are unique and we must change our ways. Not masters of war but masters of peace. And if we can only do that then we can have once again a Golden Age in this world like we had once before. Before we became what we are now.

The gods that are spoken of in olden times, and so forth, they consisted, you know, of having great flying devices, and the power and so forth. But like gods are still around. They live in a controlled environment like an incubator where everything is ideal. The temperature is ideal. War is not even practiced now. They have set up a system of rule by type[s] of biological robots who have the power to destroy if the rules that have been put together by their law are broken. Therefore, there can be no system of political corruption of any kind because they have put themselves into the hands of these genetic-engineered biological robots. They are—you might say in a sense—they are a type of computer, we don't know. They are not the masters or the rulers. They are the stabilizers in the order of all things, hence peace. This is what we are talking about, the inner realms. This is a system that we don't have.

Dan: What is the work of your order doing in conjunction with the inner realms? What do the inner realms expect us to do with it?

X: They are trying to turn out, frankly, a noble breed of [humans] who are pure enough [and] who can make contact with the inner realms and ask them to please help us; the surface world is out of control. We need [their] technology and need [their] help. And we do not want [their] help to aid war but to bring peace. This is what it's about, you see.

Now, I'll tell you something else. There was a small undersea boat built. I can't tell you all about it. It was built in three sections, fabricated in Germany, Norway, and one other country in the area of Ciara Del Fuago in South America. We assembled it and two of our most humble Heilege Ritters [Holy Knights] were the crew. They proceeded to a certain longitude and latitude and they made the dive. They did not return because the inner order says that we are still not ready, even though [they were] impressed with these Ritters that were sent. Nonetheless, they are in a state now of Paradise. They expect to live for hundreds of years. They can't return, because we're just not quite ready. In the future, I don't know what they will decide. But we think once more after a period of time [...] there is certain information now in the hands of the order [...] they will try again. It is the theory that if they send enough delegations of pure men with unbiased motives [...] men of peace [...] that it may be possible to get help to stabilize. Because they have power and can control the weather. They can do that. They can absolutely use forces that we can't understand or comprehend. It would mean that our weapons would be chaff in the wind, useless against them. Because these are the masters of the inner world and that is the reason they have set this environment up. Without it, surface man had great destruction with all the fables, fantasy stories, whatever they were, of Lemuria and Atlantis. [Humans] destroyed themselves and nearly the whole world. They let the survivors struggle because they said, perhaps if they had learned now the futility of what they had done, that they could become peaceful in the outer world. Eventually, they would understand the futility of war and destruction and all of these things.

We are back now and again approaching the time where another cataclysm will have to take place in order to put us in our place once more. And we don't want this to happen [...]. When you really think about human beings, except for this nature of war [...]. Have you ever thought about what a wonderful thing a human being is? All of the things we can do?

Dan: It's endless.

X: It's endless—music and science. Science and music is something [...]. It is the highest and one of the noblest aspects of humankind and the ability to make things. In our time, shall we say, a television set here. It's miraculous. It circles the globe in a twinkling of an eye. What one man says ten thousand miles away can be received an image and voice in a split second. Yes. All of these things are minor things we can do. But what we have compared to the Arrianni of the inner realms is like the dark ages. And it only stands to reason that we can't be on an equal basis with a God race. That's what they are, a God race. The knowledge, the power that they have is so staggering. It would be like trying to explain it to a barbarian when barbarians overran Rome. Statues, paintings, mosaic tiles, and all were destroyed just out of sheer ignorance. We couldn't compare to them. We would be barbarians. We have faced always violence, anger. We are jealous. We are imperfect and, therefore, they reject what we are. This is the idea behind the Order, a secret organization, yes, in many ways. The motives are pure. Two of our most exalted Heilege Ritters are gone, in the little U-boat. The U-boat was named and christened The Omega. But there will be maybe another Omega, and [it will] once more try but not maybe for many years.

Dan: I have a series of questions that have been going through my mind ever since we started our communication. What I would like to do is to give you a copy of the questions and then, while I'm reading your manuscript, you can kind of check off questions, many of which you've already answered here. There are different categories for my questions. I've gathered them from various books and questions from other people who are close to us. Last year, in fact, I think it was on Easter Day that I saw on TV a story about the sacred lance.

X: Really, the Heilege Lanze?

Dan: I went back and I looked at the old TV guides and I couldn't find it. But I will find it. It was in April and I'm sure it was either on Easter Day or week or two before or after Easter.

X: Hmmm.

Dan: I can't remember the exact time it was. But it was fascinating program. In fact, the woman who was helping me with my research was there with me. For lack of finding something better on the TV we turned right to it, and it was absolutely fascinating. It talked about the period of history from the crucifixion of Christ to the present day. It spoke about the lance being in the United States for a while, and then Eisenhower gave it back to Austria.

X: It's the copy [of the lance].

Dan: A copy?

X: Just a copy. And we feel that the lance is responsible for all that has really occurred in the way of good and evil. It is a most powerful and wonderful thing. It's not really completely understood what this power is, but it is there. And it is a power of light and good. But it also can be used for evil. That's about the best way I can phrase that.

Now, do you think we should go and have a sandwich or something?

Dan: Yes. [We drove to a small grocery store and picked up a couple of sandwiches. Ritter von X continued his explanation.]

X: Together we can go over this very thoroughly. We could go into all hours of the night and discuss, because you have traveled a very long way. And whatever you want, I will try to answer you the best way I can.

Dan: That's fair enough and thank you. I remember, when I was around fifteen years old, my oldest brother, Fred, told me this story. He said in a fearful but assuring manner that he was the person in a past life that held the spear and pierced the side of Jesus. He frightened me especially when he held a small-caliber pistol to his head while lying upon the floor. He controlled himself and withdrew from that moment of self-destruction just to tell me his story. I believed him.

X: So he thinks he was Longinus?

Dan: Yes. I didn't know who Longinus was or anything about the spear.

X: How did you find us?

Dan: The research work that just took place in the library. Raymond Bernard's book, The Hollow Earth led me to various avenues, and I just started checking things off the shelf and sorting through stuff. I experienced many mystical things at that time.

X: The mysterious Dr. Bernard. He founded a colony in Santa Catarina, in Brazil, some years ago in the 1920s. He explored some mysterious city back in the Brazilian jungles, and he had his information from Tibet about the center of the world. The search ended, he disappeared, and was never heard from again. I believe natives or something killed him. But there is another side to the story: he completed his mission and could not return. So in what you are [searching] for you can go to a certain point and from that point you have a point of no return.

Admiral Richard E. Byrd was also a unique person, spiritually motivated. Undoubtedly he had some contact when he was in the isolated regions, and when he had made some type of contact, he learned the truth. He made an attempt to reach a mysterious land beyond the South Pole. He evidently was successful because later he was totally discredited by the Walter Reed Naval Hospital. They called it natural fatigue. It wasn't.

Dan: What's the purpose? How much do you think the US government is involved in exposing it, yet having it covered up at the same time? Are they selecting people themselves? I've had another remote viewing experience. I was standing in Arlington, Virginia, and I was talking to a bunch of people who had a school going on and they were learning the language and studying all kinds of things about aliens. And my other research partner, she had a contact, a message came to her through a dream that the US was conducting secret excursions by the military, going back and forth to the inner realms. Is all that possible?

X: They could go one way but they could never come back.

Dan: I see.

X: If they were even allowed to enter, because—

Dan: Who do you think they would send back? I'm sure they're probably planning that someday, and if they do, what do you think will be their purpose in sending an individual back?

X: It would have to be someone who has a very normal life existence, who was perhaps even a respected man, a professional man—whatever. And then come back to society and say, "I have been inside of the Earth." I have seen it. It's real. It does exist." And put him completely on the line of credibility. His colleagues would have to say, "He's not insane." There must be something really true because he has everything in the world. Why should he place himself at the point of ridicule, and for what? [To] tell [us] we all have to change. We cannot continue the way we live and the way we are. That sounds so easy to say. However, it's not easy to do.

Dan: Sounds as if we need to let everything we know go.

X: Exactly. [And those inner Earth beings would] have to give it all up, and they [would] have to be trained how to live totally different from what they have ever imagined. Because they have forces, powers, energies, science, political systems that surpass anything we have. We have not even one good government on this whole Earth—not one that is any good. And yet, they have devices that scan man's mind. They can really see a human being for what they are. We need only to realize this. How would you feel to be in a room with a being that could read your thoughts? Would you be uncomfortable, or would you suddenly decide this is something totally different—"I must learn from them." What would you tell them?

Dan: Yes, fearful yet curious enough to want to know much more.

X: These are the types of things needed to be accepted and learned.

Dan: By accepting and learning, the consequences are probably my own fate.

X: Absolutely. It would very much affect your own fate. There's no question there.

Dan: When I encountered this being, it's a fearful kind of thing, but inviting—always inviting. But my intellect seems to be a stone in the way. You know. I feel like I have to stop in space and time—

X: I understand.

Dan: —a little more than I dare to communicate with anyone.

X: I understand what you're saying.

Dan: That's good—I feel you know my thoughts.

X: Here is this prepared manuscript. And this is a book jacket design—how it would look. It has no bearing on Nazis or that sort of thing. In there is a lot of classified information. [He removed an object from his briefcase.] This belonged to Karl Haushofer. There is some ancient Sanskrit on there that is a very old, a very, very old plaque. I wasn't aware that Sanskrit was a language in that part of the world. But it is. We had a Smithsonian expert translate it. Do you know what it says? Can you guess? Why don't you guess? [He paused but I said nothing.] It was about five hundred meters down a shaft in the Potala Palace where this came from, from the information we have. It hung over an entranceway that glowed from some type of energy. Dr. Haushofer describes it as being luminous in quality, and tingling to the touch. So we knew it was not any length of luminosity, it was energy. It hung there in that secret location. But one of the priests told him the translation. It reads, "This path leads to Aghartha." That's in Sanskrit.

Dan: Was this in 1945?

X: Thereabout. These swastikas on there are an ancient symbol and have nothing to do with the Nazis. I am sure you are aware of that.

Dan: Yes, I am. Even in Vietnam I noticed many Vietnamese wore a solid gold chain with a swastika on it. I have a photo of a large an ancient statue of Buddha with a swastika placed at chest level.

X: Anyway, this is very old. Karl Haushofer was allowed to bring it back. Part of the secret knowledge this man had is now in our Order. I won't be able to keep this too long.

Does holding it give you any strange feeling? Do you notice anything different when you're handling that?

Dan: I'm somewhat frightened. Is this written from right to left?

X: I believe you are right. Now, look at the shape of this thing itself. A Flugelrad is viewed as a disk from the side, as many of them have been reported, sometimes a strange cigar-shaped craft, something of this nature. Now, this is not just a coincidence.

Dan: But history tells us that Karl Haushofer took his own life?

X: He believed he would return to the realms of the inner world and be refurbished.

Dan: Please tell me about the International Society for a Complete Earth—is it the same order you speak of? Or is it an Order behind an Order, or are these two different orders?

X: Yes, the Order is behind the International Society for a Complete Earth. They are behind us. As I mentioned to you on the phone, we are seeking world peace, and are definitely part of the Order. We are devoted to peace—world peace. Because of the knowledge we know about the secret existence about this race of beings in the Earth, it is not fantasy, it is real—absolute. There's no question of it. One day the nations of the surface world are going to find this out, in a way that they won't like it.

Here's something that I wish for you to analyze. It's a little present for you. It is from the inner realms.

Dan: Is that right?

X: It's a geode of a very special type. It has crystal inside. If you should cut it with a diamond saw, inside are sparkling beautiful little crystals. To someone not knowing what it is, it's a curious pebble.

Dan: How did it get here?

X: It got here by Flugelrad, that's how. There are some others. [He shows them.] Like that one, see? See the little crystals inside.

Dan: Yes. I can see them sparkling.

X: Now, take the geode in your hand here and shake it by your ear.

Dan: Yes.

X: In there are tiny little diamond like crystals. That's what you hear. You see. This is an example of what I told you. The Earth was born a gigantic bubble, gas temperature, heated to where?—we don't even know how hot this can be. Cooling in space as it tumbled in space, gathering bits of space dust, cooling, and then you have an exterior and an interior.

Dan: [Sighs.] And we live on the grosser side of the stone.

X: And there you have the range of the mountains and valleys of the Earth in miniature.

Dan: This reminds me of the mystical contact I had in Iceland. Wow. How were the Flugelrads—how were these found?

X: They were delivered to one of our Ritters as little tokens. You can find, very likely, some geological expert and have him analyze them. He will probably tell you that they are interesting. "How did you get this, and where did it come from?" That little one is for you. You may have it.

Dan: Thank you, I appreciate that.

X: And if you do desire to have it cut, use a professional diamond saw to cut it. And as they begin to penetrate it, be careful of the little diamonds that will come out, as you will want to keep those.

Dan: Those are actually diamonds?

X: They are a diamond crystal of some type. They are quite different from what we know. You will find this out. And the reason I brought you these things is to give you proof that what I am telling you is the truth. You are going to find [...]. Your geologist is going to tell you, this is something quite different. [He will say] "Where did this come from?" "Where did you get it?"

Dan: And then what will I tell them? I found it in my backyard?

X: Then you tell them, "A friend gave it to me. I don't know where it came from."

Dan: Okay. What verification do you have that UFOs are a well-known fact and that they are coming and going at certain entrance points?

X: Many people could verify the existence of them, including the ones who know more about it than anyone else. That is the United States government, the Soviet government, and the English government. They can verify beyond your wildest imaginations the existence of these. But they will not. They would realize a major power struggle. We would lose control.

Dan: It's probably shelved and filed away somewhere.

X: Yes, of course. It's buried very deeply in concrete bunkers under certain areas in Virginia—all of that secret information.

Dan: Did you say Virginia?

X: Virginia. So again you encounter the earthly power of the [confidential name], who are very well aware of all these things, as far as they understand. But they are carrying the seeds of their own destruction, by the very fact they attempt to conceal, discredit, and to deceive. The thing that I would point out there is this: every individual who has been contacted is brought to some realization then suddenly finds more than he had ever thought of. He met face-to-face with some strange craft, some different kinds of beings, and it's obvious that they are superior in every way. I know this. I realized that I was just waking up. There's something going on here.

"Nobody can tell me now," they would say, and I have listened to at least three men who said this. They said, "My God, no one could ever tell me that these things don't exist. I came face-to-face on a farm road late one night, my engine quit, my lights went out, everything was wrong, and I was scared to death, and I saw these beings get out. They had on silvery garments. They looked like human beings. And then I don't remember. They next thing I know I am driving along in my car on the highway. And suddenly I begin to focus on the road, and, what's happened to me? Where am I?"

Dan: You mentioned yesterday that you would put me in contact with someone.

X: Yes, I can put you in contact with a man that has had a very great experience and not very long ago. And he could certainly tell you about his encounter.

Dan: This would help me to understand without fear and find some credibility as well.

X: He is one of the few people that I know that will speak out without fear and ridicule, because once again, he is another one who has been there, and that has changed my life, and my thoughts and my fear of retaliation.

The Arrianni are in Flugelrads. They see everything that goes on and as Haushofer has written in one of his most secret diaries, "There is nothing on Earth the king of the world does not know." So these humans sit in these buildings and think and do nothing but look out the window all day—and they are being paid very attractive salaries. The American people are not being told about this because it would be so irrational to say this man sits here in his little building and looks out the window all day long and they pay him thousands a year to do that. What would Congress do with such information? That's why the CIA and the KGB are so frightfully secretive in their work. They do work that would be considered a cult beyond reason—ridiculous, yet it isn't.

This is one thing—men like Thal have passed on information to certain Earth men and that is the only way we are getting help—because they are helping to prevent Armageddon from occurring. That is why I know that it will not occur—because there is a superior mentality involved in all of this. We are on the outside, they are on the inside, and when I say that, [I mean] they are truly on the inside of all things, because nothing can be hidden from them. So our finest intelligence is not so-called agents but people who do nothing but look out the window all day and all night. And they get information before them always on a pad of paper and pencils. They write down every thought that comes into their minds. And then it is all sifted and analyzed, and strangely enough, a pattern comes from it. Here is a message. What does it mean? How did they get it? Now we have to check it to see if it is true. So the satellites are sent, they are photographed, and sure enough it turns out to be so. And some of these men think this is the power of God. But it isn't, in that sense. Because what we are dealing with, as I said, is almost beyond our comprehension. It is so powerful. We have this, and we know it's true. There is no way that it can be otherwise but what we know.

Dan: Thal is a mediator between the Order and the—

X: He's a friend.

Dan: A friend of the Order?

X: A friend of humankind, if you want to place it that way.

Dan: How does the legend of the spear relate between Thal and man's destiny?

X: This is a strange thing. Of Christ, we believe that the Bible is a combination of ancient, archaic record, written in the language of that time, as it was understood in that time. And we believe that this man, Christ, was a Prince that came from the inner world to the outer world because of the great knowledge He had that astounded humankind. You know, it is written how He confounded even the most learned and wise at that time. These men, these priests in the temples were astounded when they heard this young boy speak, and what He knew. You know this, according to the Bible. It says Jesus died, then resurrected [and] he was the Son of God. He also contained in his body a magnetic power to refurbish himself. These types of beings cannot be killed. And this is the resurrection. He rose, and this great rock from the tomb, weighing many tons, was rolled away by what some soldiers in their presence said were angels.

They were beings—glowing beings. They had come from the inner realms to rescue Jesus. It was a magnanimous gesture by one of the great beings of the inner realms who came forward to try to help the world. And since that time there has not been one, not one bit of help. Not in the sense this teacher came forward. We believe that He was a member of the inner realms who came to help—a type of prince—and he spoke of such strange things—"in my Father's house are many mansions." He had a vast working knowledge of astronomy, of religion of its time, of politics. He had a grasp uncommon to an ordinary human and the ability to lay hands upon people and literally raise them from the dead, because they possess magnetic energy. They can do anything, and we don't know how it works. We can't, we can't comprehend. But it is there.

Thal wears a dark blue mesh type of garment. It is tight fitting and he is well over six feet. A beautiful glorified body, like no human can have this kind of body. You cannot touch him. He warns you of this great danger because there's a vibration of light all around him, and the spacecraft that he comes in, it isn't anything that a man can comprehend. You think that the aircraft you arrived in today was fantastic and beautiful. But it is a metal, riveted, put together with our technology. But the Flugelrads are something unbelievable.

The controls inside, I have never seen anything like them. There are no levers, no wheels, nothing of that nature. Everything is a rose-colored panel of light. And to get things to work is merely to pass the hand in certain directions, to say certain things. There are no doors on these crafts—not like open this way [he gestures], or a hatchway like on a U-boat, or anything. When they land, there is nothing you can see until this pulsating light stops, a voice speaks forward to you to stay clear until the lights on the outside have ceased. And then all of a sudden there's a door—not a door, it's an opening. It defies any technological explanation. These crafts are not large. Some of them might be but—

Dan: You have seen or experienced one?

X: Yes, I have seen and have been aboard one, in the countryside. I won't say where. But I thought that night we went there, I thought, hah, this will be something very funny, something that I will ridicule this person about. It was very early in the morning, heavy dew on the grass, on the meadow, with lots of stars. It was a warm summer in August and he said that we have not long to wait. And then in the distance I saw this little tiny orange light sitting up there. And it got closer and closer, and as it came down, it became huge but it was not a bright light of any kind. It was a very soft orange glow, and it—frankly, to place it mildly—it scared the hell out of me. I was ready to run, and I'm not a man who frightens easily. But that frightened me, scared me, terribly. And the dog that I had with me ran away, quickly. He was terrified. We found him two days later in the woods, scared to death of something. Anyway, the craft landed and spoke perfect German dialect, a very high type. This magnificent being told us to come aboard, [so] we entered, and it was not a bright burning light in there, but a soft light. It didn't hurt the eyes. I've never seen anything quite like it. But we were able to see everything perfectly without the harsh glare of illumination of this type. The first thing that I am seeing is no instrumentation, no controls. Nothing I could even recognize. It was so fantastic in nature that I can say only one thing to myself: "I'm dreaming. This is not real. I am dreaming and this is not real."

And Thal turned and he said, "You are fully awake and I have a greeting from Dr. Franz Philip."

And I said, "Franz is dead."

"No, he isn't. He sent you greetings." And he said something then, that I knew it was really from Franz because it was something very personal to me. I told no one about that for a very long time, and I don't tell anybody, anyone, any people, that I am immediately associated with in everyday life.

We are here talking now, [and] I have told you. Not only that, but several other Ritters have been contacted by these forces. For one thing, all the terrible weather that we have been having has been occurring in certain places. I wrote one of my friends a few months ago and told them some of the problems they would be having there. Also, some of the people in New Mexico and Arizona region—I told them to be careful of the weather there. Great floods like never before are going to hit there. And things of this nature are trying to tell people to stop tampering with the delicate balance of the Earth, because what is affected on the outside also can affect the Arrianni, [because] they do use the air as we do. They do have hearts. The organs are similar but there is a difference in the way the magnetic flow in the body, which is—I can't explain it. It's something magnificent and terrible and it's even kind of frightening. But they are trying to tell us to stop tampering with our environment. Because, you see, this Earth itself, actually, it is already is nothing but a gigantic Flugelrad itself. It's a spaceship. It was a world we knew once before but many, many years ago, when we had a perfect external environment on the surface. But then some catastrophe occurred to break this gigantic canopy above the Earth and the moisture that surrounded the Earth to protect us from harmful radiations.

Dan: It's written in Genesis, when it talks about the waters above and the surface below.

X: This is what happened. Part of the screen mechanism here fell. The Bible says this is what happened. Only this is not any information just from the Bible. It was told to me and other members of our Order what had really occurred. And I have wondered why this information hasn't been put down as absolute truth in present-day language for everyone to understand in a responsible manner, where you don't have any mystical parables to ponder. It would be like reading a scientific journal, in fact. But it hasn't been done.

Creation [and] many things that are in the Bible was written in archaic terms of those times by scribes as they saw those things. This was the Christ. We are fully convinced from what evidence [has been revealed] from the inner realms, that He was a master, teacher, and a Prince. And He came forth on His own will and volition to help us because somehow, He admired us or found something worth saving in us. And then, being the animal-type creatures that we are still, even then, we were then, we still are, and we turned on this individual and tried to kill Him. Well, He didn't die, He simply returned to where He came from, using his own magnetic power to refurbish His wounded body. And it's even like it said in the Bible—He rose, and what they said was, He ascended into the air.

Dan: They said He went into the heart of the Earth.

X: Yes, maybe that's what it said. I'm not familiar. What happened was He used this great power. As I say, they can't be killed. There isn't any way you can kill them—that we know of. And they have banished war and destruction from their own realm.

Dan: Well, that's the kind of memory that we have to tap into then. They did it. We can do it.

X: That's the point I'm trying to say.

Dan: We're just the tail end of the dog here. They're the head and we are the tail.

X: This is right, and we have to find a way to convince them that we are still worth consideration. They tried to help us one time. We didn't, we didn't take it that way. And that's what we are all about—trying to reestablish an Order, where we can send emissaries that will simply be able to approach them, and say, "Here, here we have a small group that are worth consideration." We will send one or more of our kind among them and teach them what we know—or what we want them to know. They certainly will never give power—full power to us. Because we—

Dan: We couldn't handle it.

X: We couldn't handle it. Yes, that's right. Well, Daniel. I think it's about time for me to go. I hope you enjoy the manuscript, and maybe you'll see some things there. And I'll try and be here in the morning sometime, around nine. We'll discuss some more. I'm afraid I've done most of the talking. Have a nice pleasant sleep tonight. Get all rested up. Good night.

Chapter Eight

~ CONVERSATIONS WITH ~ RITTER VON X

DAY 2

The most beautiful and most profound experience is the sensation of the mystical. It is the sower of all true science. He to whom this emotion is a stranger, who can no longer wonder and stand rapt in awe, is as good as dead. To know that what is impenetrable to us really exists, manifesting itself as the highest wisdom and the most radiant beauty which our dull faculties can comprehend only in their primitive forms—this knowledge, this feeling is at the center of all religiousness.

—Albert Einstein, 1879–1955

X: The Creator made everything into a design. [Look at] the atomic world, electrons, neutrons, etc., all moving in various little configurations and directions, in perfect harmony without any collision. A grain of salt looks like a lot of these dots all put into a certain geometric pattern, like a snowflake. It is even beautiful. The snowflake itself is beautiful. Did you ever see one? It's magnificent. There we have the intelligence of the Divine Creator in all matter, energy, and so on. We have only the works now surviving. We have to work with what we have left.

Dan: Where are the latitudes and longitudes of the gateways you mentioned on the ocean floor?

X: I can't tell you that. It's closed information.

Jan: Are there still access places without interferences?

X: No. There is a certain crystal that must be taken that will pick up the very outer fringes of the magnetic mainstream, and deflect and send back a signal and then certain forces will deactivate the maelstrom and allow you to enter. They know your purpose and your thoughts. There's nothing you can conceal from them, and this is the most embarrassing thing that you can imagine—being face-to-face with this being, like Thal. At first it was irritating, because I felt I was not in control now. I was not in control because they could read my every thought. What if I should think, that you are a very strange-looking person? What an odd garment you have there. What a strange device you have on your side. What does this little design mean? To have someone just to look at you and to tell you before you can even gather these thoughts together and transform them into audio sound, to say to you, "This is such-and-such, so-and-so, the purpose of this is such"—it's frightening. It's irritating. At first it was, because you have no longer any control. You are being controlled.

Everyone in our world has the privilege of keeping inside of our brain, so to speak, all the things we think. How would you like to be face-to-face with a being that could anticipate and listen to every secret inner thought you have? It would be like living inside of a goldfish bowl. And this is the capability of these beings. Therefore, when the Omega approached, they had the crystalline material and the transmission took place. And the Ritters knew when they were allowed to enter. If this was not successful, they could not return. So, therefore, when these two were sent, they were specially chosen. They were men with little family ties who were dedicated to purpose, only to succeed without regard to self or circumstance, whether or not they would be allowed to return. But also [...] they would be living in a type of Paradise, and would live hundreds and hundreds of years into the future, and one day could become as the superior beings of the Arrianni. What a trade—the prospect of a type of eternal life compared to releasing yourself from a few short years here on Earth. This is what they were faced with. They chose the mission gladly.

Dan: Who wouldn't?

X: So they know your thoughts, and if you have any impurities, you can't hide them. Therefore, under such circumstances, you force yourself to think only pure, idealistic thoughts, and convince yourself, your inner being, that what you are thinking is true, and you mean what you say. You speak without deception. Deception of the Arrianni is impossible, absolutely impossible. This is very difficult for man to understand, let alone accept it.

For instance, Daniel, say you were married to a beautiful young lady of Arrianni extraction, who had this extreme, marvelous power. Do you think you would really be comfortable with her under your present circumstances? Being an ordinary human being, knowing that she was an extraordinary being? Do you think you could live with someone like that, who would know your every thought?

Dan: It certainly would humble whatever—

X: Humble. It would humble you, yes. It is humbling indeed.

Dan: Absolute humbling.

X: I don't know whether you have ever talked to the contactees much, but they say about the same thing.

Dan: No, I haven't. That's a new area for me. I didn't want it to become like an Alcoholics Anonymous group just speaking out loud to a crowd, or especially to a clinical psychiatrist. It's important for me to come in contact with myself and continue the investigation and research, and find answers to questions and gain knowledge of what's going on.

X: Yes, there are people who falsely claim they have had "close encounters." Some people who claim to be abductees can only be publicity seekers. They speak in some way about craft landings and all these weird things. A door opened and there was a ramp that came down and had some legs on it and out came a four-armed green monster with a big glowing eye. It is not so. It's false. It didn't happen. It never was. These people want to get their name in some papers—like one of the sensational tabloids at the checkout counter, something of this nature, where they print all of these strange headlines, you know—"My baby was abducted by a UFO for two years" and things like that. These people are looking for some sensation for themselves—to see their name in print, and I guess that is their reward.

Dan: Yes and therein is their reward.

X: That's sad. And what must their friends and neighbors think of them? They are totally discredited—ruined. But those that have really had true real encounters, is a much different matter. They shun any kind of publicity, but they generally are so convinced of the truth that they don't care what anyone else thinks. The truth will set them free, once again.

Dan: Well, that's something that nobody can take away.

X: That's right.

Dan: Okay. How is the nuclear testing affecting the Arrianni? There are massive demonstrations going on in Germany and in everywhere in the world. It's getting really out of hand and people are—are—

X: Fearful, because they realize this rampant energy turned loose on the world can affect so many things. You see, everything that we have in the world is a type of electronics, or even more truly, magnetic energy. Gravitational forces of magnetism hold us together, and so with the influx of the erratic particles of atomic energy, it is like the same thing. The world is a body, as we are a body. And as long as our blood is circulating at a rate with normal cellular structure, we are fine. But let erratic cells become loose in our system, as a cancer cell does, and it results in a change of the physiology of the body and its functions, and in the end, destruction. This is what the Arrianni are concerned will happen, as it has happened before. This is the fourth and last world. And I don't know what that totally means. But this time it will be different. Does that answer you?

Dan: Yes. What is your knowledge concerning their weapons, machines, etc? And why do they have weapons?

X: Weapons are still in existence only for sojourns to the outer world. They don't need weapons inside. As I have told you, they have this perfect governing system of robotoids, so that their weapons are not necessary. War is out. But the outer realm of our world remains hostile. Let us say we have in a jungle a dwelling that is a fortress, completely impregnable. Inside is food, water, everything we need. We are happy in there. We feel secure in there. We carry no weapons in there. All of this is good. But also we know when we go out and "open the door" and go forward into the jungle, maybe a few kilometers, we must have weapons because there are ravenous beasts out there who may attack us at any moment. Therefore, we carry weapons. But not in our own homes because we know we are in an enclosed, safe environment, where there is no need for weapons. But the weapons that they have are so terrible that they can take the flesh off of a man, a human being, while he still stands on his feet. It's magnetic energy. It's a power that the most insane warlord on the planet Earth would probably murder his own mother to obtain, because it would be so valuable, it would satisfy their megalomania, delusions of power. In other words, if I had such power, I would be God. No one could deny me. But they don't [use their power in that way]. That terrible power is not used indiscriminately.

Dan: We don't have it?

X: That's right, we don't have it, but not that men don't desire it.

Dan: Desire runs rampant.

X: The Pentagon. You should write to the Inventors' Bureau, and find out. They will send you a brochure that states very emphatically that if you are an inventor, here are some things they need [invented]. Among the things they need is a death ray. They need a type of gas that will paralyze every living thing. Then they will tell you at the bottom that this is all promoting the interest of peace.

Yes. All of these things they want, death rays, and so on. You can get a list of this by simply writing them.

Dan: Is that right?

X: So the desire is there. What we are saying is this. The Pentagon is saying to the inventing young minds of the world, we need your knowledge and intelligence.

Dan: So we can blow ourselves to pieces.

X: So we can do just that.

Dan: It reminds me of the inscription that's inscribed on the doors of the United Nations building. It's something out of Isaiah that says: "And they shall bend their swords into plowshares." The scripture gives me something to imagine.

X: I believe the philosophy now is to melt plowshares into guns.

Dan: Do the Arrianni fight among themselves or with people from other planets?

X: No. No. They have perfect peace and harmony. Thal has told me that the inner realm is one vast garden of indescribable beauty. And that there are marvelous structures there, great halls of learning. He said the noblest aspect of life is to learn the arts, the love of music. And if you can imagine, even the harp is there—the beautiful harp. There is nothing more beautiful to me in music in the world, strangely enough, than the harp. I love the sound of it.

Dan: Oh, yes, especially when the acoustics are just right. It is vibration of indescribable beauty.

X: It's so odd that it does tell in the Bible of the angels carrying harps and so on. Among the Jewish race is a great love of the harp. I was most impressed to know about the Jewish people, [and] to learn about our culture and love of our family and to play the harp well. There's no strife, no war. The greatest love is the pursuit of true knowledge. It is true knowledge.

Dan: Referring to The Smokey God by Willis George Emerson, it says he left maps, diagrams, etc., of the inner realm which were later donated to the Smithsonian Institute, but as we discussed yesterday, they are not there. Where would I expect to find such information?

X: [He seemed to avoid the answer.] Naturally, it won't be there. But what he wrote had a lot of truth in it, and I think those beings that he encountered were the last of the great race that was once on the Earth, in the garden times. These were men and women beings of great stature and beauty. I think the Bible speaks of them as men of renown, when giants were on the Earth. They belonged to another time, when things were vastly different. Only occasionally, genetically, somehow do we have giants born. And about the last one I think that the American race had was the one called the Alton Giant. He was a gigantic man nearly eight feet tall, wearing large shoes. They live a very short time, but they did exist at one time.

Dan: Yes, Genesis 6, Verse 4 reads: "There were giants in the Earth in those days; and also after that, when the sons of God came in unto the daughters of men, and they bore children," etc. Does this have a connection with the sightings in Northern California, or Bigfoot or Sasquatch? Sometimes it seems like it's one of these beings running scared from humanity, hidden in the forests in these remote places.

X: Professor Haushofer told in his writings of the time he spent in those remote areas in Tibet. They have a name for them there. They call them Yeti, I think. High mountain climbers have seen these Yetis. The Tibetans regard these creatures as very old. Not dangerous. The professor describes them as a genetic mutation, probably existing from the time of the great cataclysm when the basic life form was altered. They are a type of man who had his genetic code altered and thus adapted to the cold high places with fur and all of these things. But they remained obscure, gentle creatures that hurt no one. He calls them children of the woodlands and the mountains. They may live inside caves, and so forth. But they only venture out when desperate for food. It's not really known exactly what they are, but they do exist. Some could tell you that the Yetis are a kind of God. But it is my own conclusion that these are primitive people [the Tibetans] meeting with [a mysterious creature] and not understanding it, [so it] becomes of vast importance, perhaps a deity of some type.

But the scientific mind of a German—Haushofer—saw them as a genetic mutation that survived a cataclysm of some type. They are always found in the high, lofty places. That suggests that the Earth was inundated and in a state of cataclysm, and they went to the highest place where the God force, or whatever it is that controls, [sets adaptation into motion]. Do you know we ourselves, if placed in an environment hostile to us, many of us probably would die? But there will be a small nuclei that would adapt if it were bitter cold and there were no animals, where we would naturally [develop] their skins to keep ourselves warm. We would probably grow long hair on our bodies all over as a protection against the cold. And if we encountered some other races that had survived in another remote part of the world after a cataclysm—[a part of the world that] was warm and nice enough, radically different from the temperature of ours—if we came face-to-face, we would have the meeting of a human and human. But [these humans] would be animalistic in appearance because of what nature had given them to survive. [They would be] of basic animal intelligence with fur and hair, [because they had] survived that way for thousands of years.

Dan: We can always become Yetis.

X: [Ritter von X laughs] let's go and have a nice dinner. I want us to go down to Long John Silver's today and have us some shrimp dinner. You like shrimp?

Dan: Yes, that sounds excellent.

[While at lunch, Ritter von X began to explain how the Arrianni could be contacted, and how we could have them land their Flugelrads, and how to keep it all confidential. We continued the private conversation during our hour-long lunch, and then headed back to the meeting place.]

X: They are capable of such technology that it's mind-boggling. So the best thing that I have been told is to accept it and understand it. In time you will.

Dan: Time?

X: That's another relative factor.

Dan: Well, that's our life.

X: Time means nothing to them. That's the difference between immortality and mortality, which they possess the power of.

You yourself are no ordinary man, or you would by now have said what you probably thought—that it is so terrible that this guy must be totally crazy.

Dan: Well, I've learned not to voice certain things. I've learned that the hard way during my second marriage and most of my life. I have had so many mystical encounters and many different experiences. I have devoted myself to the truth. And what I asked for I did receive, so now I'm careful in what I'm asking because it requires lots of changes.

X: Adjustments, yes, very much.

Dan: I have upset a few people because of this.

X: Upset them? Yes, because it's so different, eh? Understood, yes it's understood.

Dan: So I found a new tolerance. It's a new education in every way. The only important thing is to learn how to communicate on this level. [Pointing at my forehead and referring to third eye.] So I must bridge a new understanding for communication. Would the CIA, KGB, or any others interfere with any individual upon receiving this kind of information?

X: If they considered you dangerous enough, I think, yes. I know for a fact that my own mail has been monitored many times. They are not even very clever about it. Maybe they want you to know. I don't know.

Dan: When you called, I had a real hard time hearing you. It was like half the input on the line. I mean, the phone lines are not bad. But it was at least half the signal or little less.

X: They watch us. But we have nothing to fear because we are in the right. We consider them forces of darkness.

Dan: Understood.

X: Many of our members are monitored, letters opened, undoubtedly photographed. But we have certain means of communication that they can't make too much of.

Dan: I guess it just depends on what certain individuals are getting into, what kind of information they are receiving. I guess that they wouldn't want to really get involved.

X: Well, they go about it in different ways. One of our members of the Order was approached one evening in a casual way, [by] a stranger, you know, and he began to talk. Right away [our member] could understand this man was strange about his questioning and so forth. No doubt the stranger was very clever and was purposely maneuvering around to interesting subjects and from then it became a ladder—a little bit this way and a little bit that way. He gets you right where he wants to go and talking about this and that, religion and one thing and another.

Then the stranger came out and said to him something like, "Did you ever hear of an Order that claims to have a holy sword?" or something like that. The stranger knew what he was saying. He didn't mean a holy sword. He was speaking of the lance itself. Just little things and trying to [ascertain] what you already know. Even some of the children at school have been asked questions by, I suppose, teachers who have been coached to try to find out what kind of things are being talked about at home. Children are innocent and would tell the truth.

Dan: Very clever interferences—very subtle.

X: Yes, in many ways. My last flight back to [confidential location] before I flew on down to Munich, I was sure that I was sitting with one of them. We caught the plane from New York and [confidential location]. "Where are you going? Oh, excuse me. Allow me to present myself..." I don't know the name. "I'm going to Germany. They make good stuff over there. I work for this company, such." "Oh, that's nice. And where are you going? I'm going to Germany too." "Oh, we're going to the same place. Where are you going?" "My first stop is going to be in Cologne," I said. "How long are you going to be there?" he asked. "Maybe we can get together."

So we talked and we talked. He talked about the war and about a lot of things—as an American he would not really know, unless he had reasons to know. And so I didn't say so much. And we talked about a lot of things and finally we came to talk about some art treasures that disappeared and what happened to them, and things of this nature and on and on. And then he finally said something about the lance. And whatever happened to the Reich's lance. We never referred to it as the Reich's lance, in that sense.

It's the Heilege Lanze [Sacred Lance]. And he spoke about General Patton and having it in his hands for a moment. But what they had was a copy. He knew a great deal more, but no sooner than we reached [confidential location], he hands me his card from some machine tool company and [says] "I will be staying at such and such and call me and we will get together."

But I didn't contact him again because I was sure he [...] but things like this has happened from time to time. You don't imagine. You know. You simply know. They want to know your business. Where are you going and how, on an international flight? Then if you know [...] they have this idiosyncrasy of projecting on you about what they are doing, and telling you all about themselves, and then they start asking you all these questions. You know what I am talking about.

Well, definitely, the CIA and the FBI and other agencies—they very definitely monitor. One of our friends sent for a file regarding himself under the Freedom of Information Act, and he got it after a lot of difficulty and paying fees here and there, handling charges and whatever it was he did. But he found out all these things about himself from his confidential files.

Dan: Was it true?

X: Most of it. Somebody had done intelligence work and so on.

Dan: Yes, I had once requested my information from the CIA under the Freedom of Information Act. My only reason was that I suspected that I have been under investigation at various times.

X: You don't need any reason, but soon enough they will stop that and it will be discontinued and you won't be able to do it.

Dan: They wanted more specific questions about what dates, times, and so on. I was somewhat puzzled why they asked me about dates and times. How was I supposed to know? So I just forgot about it.

X: They don't make it easy for you.

Dan: Well, it's probably not necessary for me to know, and it doesn't matter.

X: All that matters is it lets you know you are being observed for no good reason.

Dan: If the destiny of the Sacred Lance is about to reveal itself in some way, I'm sure somebody might be looking over one's shoulder.

X: They are concerned, yes. Okay.

Dan: How did the origin of the name Arrianni come about?

X: They have been known by that name for centuries and centuries.

Dan: Do they speak verbally like us?

X: Yes, they do, but they have greater mental powers than anything we possess, so they treat us on a level that we can understand. In terms of level of communication, we would be compared to—if they were no doubt intercommunicating amongst themselves—we would be compared to perhaps the difference between an orangutans and a human being. There would be many things lacking, not understood. So they communicate on a conventional level, but they definitely can project into your mind and know everything you think. And also can say things to you, without ever any communication from audio or vocal sound. They don't need that. They have that ability.

I know one morning about three o'clock or thereabout I was awakened from a very deep sleep and told to go into a field [where] someone would be coming soon. I got up and got a flashlight and I expected a Flugelrad landing, but that wasn't the case. I met with a being that delivered a message to me, and then walked away and completely faded into the darkness as if he [or she] had never been.

As I say, they know where you are. They can scan your mind. There's nothing you can hide from them—absolutely nothing. If they want you, want to summon you for some reason, you'll certainly know it. It's like something has touched you, and then you hear in your mind certain requests. They only request a response from you. That's about the best I can tell you.

Dan: What were these messages about?

X: Only at a later time will I be able to relate this information to you in a confidential manner.

Dan: Ok. Are satellite photographs of the North and South poles available? Do you have any such photographs?

X: There are two or three photographs that you can obtain from the Bureau of Aeronautics and Space Administration in NASA. I think they are referred to as EE4 or EE67 [or 84 or 86] showing photographs taken by satellites over the polar areas showing definite apertures, very dark, round, a little bit oblong. But you can see from a great altitude as looking down. And you can obtain those. Now there is [...] and if they are no longer available there, I'll tell you where you can get one, if you write this down.

You may request copies of these photographs from NASA depicting the apertures of the Polar Regions. Tell them that you would like to purchase them. They will probably cost you two or three dollars for printing and mailing. They have some very excellent photographs of that.

Ray Palmer at one time published a beautiful, first-class magazine called Search. It was a well known UFO magazine published back in the early fifties. Have you never heard of it? When he was alive, was a very interesting little man. He was a crippled hunchback, but he had a great and brilliant mind. It has always been their policy to publish little known matters and the truth.

He published a lot of material that they have concerning the Hollow Earth theory. He doesn't say that it is true or says that it is not true. He only said that you need to judge for yourself. Also there is another publication about the hollow earth by Richard Shaver. Have you ever heard of him? I will send you the address where you may purchase these publications.

Dan: Okay, thanks. Yes, I have heard of him. Oh, sure.

X: About the secret world and all of these things. Richard Shaver was undoubtedly a contact, long ago. His writings were so advanced. He had terminology, descriptions of things that are now just becoming known to be true. And yet he was looked upon in those times as a strange fiction writer. Many of the features he had written all those years ago have now become reality already. Shaver was imprisoned unjustly for some of his alleged activities on the premise of these things.

If you ever encounter these dark forces, it is better to tell them that you are a researcher, a writer. You don't say whether they are true or untrue. That's a good policy. And yet when some things happen to you and you know the truth, then you know.

Dan: I can't deny that and no one can ever take it away.

X: Exactly right.

Dan: I recently had a psychic experience. It was an out-of-body experience. I felt a strange darkness within myself. And right above my inner darkness was an incredible light. At that moment if this darkness was suddenly removed from me I would be totally absorbed into that. Strangely, I felt the darkness was almost like a shield, for something I was not quite ready for yet, but it would be there, in due time. It was a magnificent feeling of great energy.

X: And you felt very good.

Dan: Yes, and I understood the darkness without the sense of fear of it. It's the light that I seek and I know it will impart all its knowledge, power and wisdom. Many of my mystical experiences are like that—I encounter a direct light. I have no form whatsoever. And luckily I have always returned. I don't know why I keep being hurled back into the body or why I'm being taken. This intense mystical vibration overcomes my whole being, and then it vanishes! I've learned to flow with it and not against it. Before I was resisting now I learn to yield to the light without fear.

X: That's it exactly. Learn to flow with it, the energy itself. You're right.

Dan: It took a while, but it's gone and now I'm relaxed. Let go and come what will. The book I'm writing is about that. It's just a way to organize my thinking. Seeing and discovering some kind of pattern of what's going on as time marches on. My life is a mystical journey.

Now, I have a question from a woman who is helping me with this research. Are there any secret US military excursions, or at least are they attempting or preparing any kind of contact with the inner realms?

X: The United States Navy learned a severe lesson in 1947. They lost vast amounts of equipment and some men were killed by a force they didn't understand. Their role now is mostly of observation.

The US Military knows the Arrianni exist and know where they are from. They can't possibly know how it is in their world. But now the military knows that they are present and that we observe them. It would be a fact of seeing images on a screen or motion picture. We could observe it, and know that those are real objects there, flying by. But we can do nothing about it.

When the US Navy met with a certain force, it evidently taught them a great lesson; also, that information no doubt has been shared with others that know. There is a channel between the Soviet Union and all these places. They do exchange information on certain subjects of mutual interest, not combatant or confrontation subjects, no. [They ask,] "What do you know about the strange objects in the latitudes of such and such?" "Have you seen them?" We have seen them, but what do we know about them? Well, what we know is that we trade information, this sort of thing. There are a lot of things I could tell you about.

Dan: Have the members of your organization been drawn together by some outside forces, such as telepathy from the inner realms for some specific purpose of mission?

X: Oh, yes. The many missions of the Ritters are to change [what we call] human nature. The numbers are very small because the numbers are only at the moment 3,019. That is a mere spark in a sea of darkness, but nonetheless it is that amount of light and it can grow. And as light displaces darkness, you have total light, and with total light [you have] absolute understanding of everything about you. This light coming into a dark room, it could be totally black and obscure, yet if you strike a match and hold it [close], you can begin to see what's about you—you see beds, furniture, and things as you look around.

The object is to keep the light and increase it. As you increase it, you replace the darkness. And when the darkness is gone, you have an environment without darkness. You understand what everything about you is, putting it in a very simple manner. When you enter a room, you have no knowledge of what it contains. But with the light you are able to ascertain what is in the room and what you see. [You] become not fearful but familiar, and you understand what it is. That's the object of the Order: to disperse the power that enforces the darkness.

In other words, you feel comfortable in a lighted room with an environment you are familiar with. And in the dark, all familiar objects are no longer there and [you lose your] sense of comprehension. You may feel that they are different, but you can't feel the color, the vibrancy. You can imagine the shape without light. But with light, you realize whatever everything truly is, and you are comfortable with [your] environment. That's what I am saying.

Dan: Do the Arrianni have the ability to control our minds? If so, do they? How long have they been doing this? How do they do it and with what purpose? Has there been a lot of activity regarding this mind control?

X: The forces of darkness like the dark operations of the CIA and other related agencies are practicing mind control. Yes, absolutely. Some of their experts have been exposed. One of the experiments took place in California, region of Oakland. The chemical department of the US Army dispersed into the atmosphere a mind-altering substance to watch the effects on people. The army has even admitted this. Do you remember when this happened?

Dan: No.

X: It was the US Army around 1963, somewhere along in there. Let me see. The dangerous mind altering drug was given to the Vietnam veterans as an experiment by the Veterans Administration. The US Army then released the chemical substance into the atmosphere in an area called Haight-Ashbury, San Francisco, wherein lived the so-called flower children and all of these which the US Army regarded as potential enemies. They released these chemicals into the atmosphere to alter, to make the people appear to be totally under the influence of strange unknown drugs. [This was] in order to attack the culture, discrediting and even attempting to destroy the genuine peace-loving people. [It was] an example of how crowd control can be achieved with these chemicals.

Only something went wrong with one of the members of this unit, who died. Before he died, he wrote all this information down and gave it to his mother, and said this must be known, what the US Army had been doing. And she brought litigations against the US Army, as I recall, and they admitted that they were doing research but it was for defense purposes of America. [And] the US government was very sorry that it happened, but some of it was somehow released with this experiment.

[Then] the mother's son tragically died, because he had absorbed too much of it accidentally. But this was a type of chemical that could be sprayed or released in some way and [in] the most innocent manner amongst demonstrators against unpopular political schemes, in order to make the crowd appear as though they were drug crazed, and to be treated harshly because of this.

Dan: I've heard some things on this, but not—Wow.

X: Some of it was conducted within the army itself. Smokers were ingested with cigarettes, [with alleged] strange chemicals in them: "Here, buddy, have a cigarette." "Thank you." And moments later, several of the chemical unit members were standing by, watching to see what happened and taking notes and recordings and all. All of this was the work of the CIA and the Army Chemical Corporation that exists right now today. These [are] strange things, and the agencies are carrying on these experiments even further, but with much more caution.

Dan: I found the International Society for a Complete Earth [ISCE] in the Santa Cruz library. I was just pulling books off the shelf, gathering information, and [I thought], "Now, let's check this, what's in here?"

X: Yes. And there it is.

Dan: There's somebody out there. So that's how that happened. I got twenty-two questions on the ISCE, in the form of "What are your organization's goals?"

X: The organization's goal is to inform the public in its own way that there is more to Earth than just what we know on the surface. There is an inside world and an outside world.

Dan: What kind of research does this organization need to accomplish?

X: When the Omega was sent, it accomplished a great deal. This type of information is fairly well classified. This is not just my doing, by any means. The research being done is controlled and classified. We only let people know the theory that the world is possibly inhabited by other beings of a different nature entirely.

We try to explain to them how this can be done by showing them perhaps a drawing of a sphere with the crust about seven hundred to a thousand miles with openings at both poles where water and air currents flow, and how even a ship sailing on the surface, like so, would sail as if going on a very flat ocean all the way around. There is a sea inside. Do you think that can be done? People in Antarctica and someone at the North Pole [...]. The person in Antarctica, the head would be here and the person down here is totally upside down. But from the gravity and the influx of power it's known that the blood does not rush to your head because of the gravitational influence. [Also, physiological and magnetic factors of the world must be taken into account.]

Dan: That's like, you can walk on the top of the surface here, but it's gravitational pull.

X: Exactly right. For example, if you had a long rod and a man walking along the top of the rod and then he went right around and on the bottom, his feet are this way, but he's upside down. He doesn't know that because the gravitational influence doesn't indicate [it in] any way. Again, the sphere with the opening, a ship could sail just like this. And this is what happened with all of the [...] sailing around. [Suddenly the phone rings in our small room where we have been discussing all these confidential matters and Ritter von X comments.]

X: So, maybe, someone wants to know if you are in here. And if you're not in here, and perhaps some agency wants to come in and perhaps go through your papers to see what you are doing.

Dan: Well, I don't have any papers here for them to look at and I certainly have nothing to hide.

X: I'm not being melodramatic, but see, I'm not surprised that maybe this could [happen]. From the time that you began to investigate all of these things and they learned, for instance, and now that you are becoming a serious inquirer, then they will begin to monitor your mail and watch you. [Knock on door.]

X: The housekeeper. We don't need any housekeeper. In this case, maybe it was only the housekeeper wondering if there was anyone here and why no answer on the telephone, eh? They will begin to take cognizance of you. The one blessing of all is that they do not have unlimited forces nor resources to watch over. Therefore, within that small framework you have the freedom of working. They are very much concerned about the increasing belief in the world now about the Earth being hollow.

Have you any idea of how many people really out there are beginning to wonder about this? There are thousands. We receive lots of mail. I can't begin to answer it because of time, and also I have many clients [whose letters] come without any return stamp, provisions of any kind. We simply can't answer all of these. So, you have all kinds of [...] believe me. I was laughing a little because of some of the very strange letters I receive. Some of them ask the most preposterous questions.

On the other hand, other inquiries [come] from many engineers, scientists, and doctors. They'll ask me if I received a letter from an M.D., a medical doctor. He is writing these books about this very subject. He believes it to be real, and he is writing these books under the name of Bormann. And he's a man who has written for scientific journals in research and medicine. But these are the people who we are interested in, mainly because of their intellectual background, and their personality, and so forth. I can tell you, what good does it do to waste time with the idly curious when you have a choice. What is more important, the idle and curious or men of true merit? Are they sincere or insincere? That is what you have to do.

Dan: How do these people come in contact with you? Do you advertise?

X: People from all over the world know about our society. We receive letters from practically every nation, including the Soviet Union. From the University of Leningrad, we had practically what amounts to a scientific inquiry, wanting to know all about how we base the idea, and what proof do we have, and all of these things. This includes the international postal coupons to pay for delivery. And so we answered them in part, twenty-five, thirty pages or so. And we found out they did receive it. They acknowledged the information to be analyzed, compared [it] to scientific credibility and theories [from] Italy, Holland, Denmark, the Scandinavian countries, and [even] from Arabia. So we get all of these things and it's a time-consuming matter. So what we have done now is that we have divided up correspondence among several of us to meet this matter.

There's a good friend of mine in Los Angeles that has worked for years with MGM. He worked for two or three years with rather famous stars. His job was just to go through fan mail and to pick out only [certain letters] here and there for replies. He told me you would learn to identify almost instinctively where it goes. After two or three lines of the letter, you knew if it goes into what you call File Thirteen, or instinctively you know what to read and respond to. A lot of it is done that way, because it's not a matter of choice. It's a matter of necessity. In other words, you account for the best you can and hope you made the right decision on this particular one.

And, of course, we value always someone who is sincere. And always we are interested in professional men who know a lot of the ways of the world. Not biased to what may be. Someone might have radical theories. Because you have to be [...] it goes back to the days of Bormann. One must be as radical as reality itself. Now what else do you have here?

Dan: Do you have a full working staff? And how is the organization financed?

X: The organization is financed on an international basis. We are provided certain funds for a year; some of this goes for equipment, stationery, correspondence, postage, and responsibility. In other words, a letter of merit would always be answered. But when you receive [...] some of these letters handwritten [...] almost indecipherable, like [...]. They would say something like, "Dear Sirs: We are interested in your organization. Can you send picture of one of these people who live inside of the Earth? Do you know how they fly in these machines? Do you have a picture? And can I ride in one if I join your society? [Laughs]

These are some of the letters we discard. We don't bother to answer them because they are ridiculous. We can't waste time on matters such as this. So we are not looking for joy riders that want to go up in a Flugelrad and talk to the inner space beings, as some of them are called. So they can go home to their little town and say, hey, I got into a Flugelrad. I met a guy who lives inside the world. Wow. What an experience. [Laughs.]

Dan: In my own circle of acquaintances back home in California, we have laughed pretty hard about things like that.

X: Yes, of course, we understand each other. Yes. I am sure of that. [Both laugh.] I will tell you one of the funniest things that I recall to mind. She was from New York. [Hilarious laughter.] This letter was addressed to us and the first word was, "Darlings. I'm twenty-six years old and I want to get it on with one of these guys that fly UFOs. I'm sincere. Please reply immediately. I promise he won't be disappointed in me." I could not believe my eyes. These are the kind of letters we don't answer. There are all kinds of kooks. [Laughs.] This is the kind of category we don't waste any time on. "Darlings," she began. Well, I was thinking to myself, "Oh, I'm sure. I would like to see a photograph. This must be a dog. She must be desperate—willing to take on an inner spaceman." There is much variation with such crazy prank letters. I don't know if they're really pranks or just lonesome people out there. I have no idea. We just don't accommodate them because—

Dan: So in time if they're sincere, as you say, they will get answered whether they correspond or not.

X: They write sometimes two and three letters and inquire that they have not heard from us, and then we do grant them a reply. Generally, we briefly answer what questions they have.

Dan: It must be hilarious to read some of these.

X: Yes, it is. I save none of these correspondences. I burn them because I have not the space to store them, for one thing.

Dan: You can write a book on a whole different category.

X: Some of them have been very funny. You get letters from children even, and I sometimes answer them. They say, "I am at [...] school and I am eleven years old. My class is writing a paper on the Hollow Earth, and we read about your organization in the Chicago Tribune," or something. Then they have all these questions about down there. "What do they look like?" "What do they eat?" "What kinds of clothes do they wear?" "What do they do?" We try to tell them in a small rational way, and we generally almost never hear from them again. They would say, Thank you for answering our class inquiries," and things of this nature. You have some interesting theories, and never hear from them again because—

Dan: Class dismissed.

X: Satisfied. You're dismissed. Exactly right.

Dan: That's interesting.

X: It is, more than you can imagine. But that was one of the funniest [...] "Darlings, I would like to get it on with an inner space guy." I told my wife about this and she said, "Well, there is no accounting for taste."

But you know, [all this goes back] a long time ago. It goes back to the Bible again about giants in the Earth in those days and men of great repute and sons of the Son. What does that mean? Who contacted the Earth and saw the daughters of men were fair and began to have children by them? These beings mixed with the human race. It makes you really wonder about how accurately the Bible was written, because it does answer a lot of questions in some roundabout ways about how all of this began and all. It really does. What else do you have there?

Dan: I'm so overwhelmed and so thankful to be in your good company.

X: I have reviewed your questions here and oddly, or not so oddly, I'd say, I found those dreams interesting, very much, and also all the questions you have. Basically, we had answered them last night before you asked, did we not?

Dan: Yes. We've answered quite a few of them. I started raising more questions this morning but I had to slow myself down.

X: You've become overwhelmed with enthusiasm.

Dan: Hopefully, I can keep both feet grounded, whenever possible.

X: It is most different in that way. We have one foot in the spiritual realm and the other foot in the material realm.

Dan: That's a tough one. This is what all this is about. Perhaps my vision of the crucifixion and the understanding of the power of the Heilege Lanze would be a gift of prophecy. I think it would be a balancing point of a great power between spirit and matter.

X: Exactly.

Dan: The great prophesy is left undone. The only way I feel that it will come to pass is when a sudden current picks me up and [moves me], and everything gets received by [the prophesy] and changed by it. I sense something way beyond my expectation. I feel I'm losing myself again.

X: You are receiving correct. The end goal and accomplishment if man can prove himself worthy is to receive the greatest gift he can imagine. It is not only a type of eternal existence, but [a] godlike existence. Not what we are now.

You will come to know the prophecy of the Heilege Lanze, which reads: "Whosoever possesses this Holy Lance and understands the powers it serves, holds in his hand the destiny of the world for good or evil."

Dan: In order to complete my journey and give it proper direction into the world, there's a lot of work I've got to do. Many changes in my mental process need to transform. I must overcome my own darkness. I've got a lot of things in my own thinking—selfish elements and things—to overcome. My God!

X: You know, the Prince [Jesus] came here a long time ago seeking to change the world and was torn because of it. But, you know, He said to the two thieves on the crosses on either side, He said that they would be now with [him] in Paradise, yet not speaking of a Heaven. Paradise is inside of the world. That is where true Paradise is. It's hard to imagine that. But that's where those beings are.

Now, even though they were condemned for some crimes that they committed against the Roman government, the Jewish government. The Prince, he saw the transformation of spirit in them and [felt] compassion [for] those people. Actually, in the end he rejected the world, the outer world. And it's been in a very bad state of struggle and world turmoil ever since. This is why we now know who He is and where He came from.

This Bible is a chaotic record written in such strange manners and ways, but He will return. And when He comes back, He will claim that lance. And when he does, it [will] fulfill a cosmic destiny, a completion for this world we live on. All things will be changed. We know this, but we can't tell or communicate this to too many. That's why it is a closed Order. The truth has to remain in a small circle. Those accepting this idea know that this is different. I am sure you will realize what is said here. It is different from anything we have here now.

The struggle is between darkness and light, because of that, I said, this will bring many inner changes to those who seek the truth. When that happens, some of the Ritters will come forward and open the door to the inner self. They are called Perfecti. They are not perfect but they are seeking perfection. That is the noblest aspect—to perfect man. To make him pure and beautiful as he once was. Not, for the most part, the degenerate creature that most of humanity is today.

So the numbers will be few, and there will be a weeding out of those who cannot [believe and accept]. The world will be like a garden again, and there won't be any struggle against thistle and thorn. This includes people who will not absolutely accept. They will be given every opportunity.

Can you think of anything any worse than people who are so blind as [to not even acknowledge] when the Prince was here and worked what were considered miracles in those times, to raise the dead? He healed the ill. People who were lame were relieved of their affliction. And yet, even though all of those around saw those fabulous works of miraculous healing of Jesus, such as [when] Lazarus comes forth—[Lazarus] was dead and then he came to life. Jesus final tribute to the new world [was] when he rose from being torn and pierced, and rolled away this gigantic stone, and the Roman soldiers outside were stunned with this brilliance, this presence, this glowing entity of energy. And then, [even though they had] witnessed all of these things, still there were those who would not accept. What can be done about it? What can you do with those people, who witnessed miraculous healing, and yet they still reject? What do you do with them?

Dan: I'm sure it's in the plan.

X: Yes, it's in the plan. As I said, the garden would be again [in] peace. There would be no deception, because all that are of the power of darkness will be destroyed. This is not the advocating of violence, but we do expect that what God had put in, to order, to be, would be carried out. But if it is not carried out, can you imagine what this world will be? It would be a dark, evil, foreboding place and would be like living in a dark cave full of reptiles.

Dan: Yes, a place where the light grows dim, until it just goes out and becomes like the other surface planets and becomes barren and desolate.

X: Exactly. Just like when the planet was void and there was no light. We could go back to it if it were ordained. But it isn't.

I seem to get deeply philosophically involved when I am talking to you. Please go ahead if you have anything to say.

Dan: I would just like to review our questions. We did touch on quite a few things.

X: Most all of them.

Dan: I totally forgot what I had in [my notes]. You have communicated with Thal. You've actually communicated with him personally?

X: Oh, yes. It's a distinguished divine privilege. He's a friend to the world yet limited in what he can do. He can only suggest and never coerce, force, or threaten. Like any master who is carrying a mantle of light. He can't do it for us.

Dan: Yes. Exactly right. He cannot do it for us. If he did, it might be violating cosmic order.

X: It would be, absolutely. You see, through it all, we have freedom of choice. We make our own prison, our own hell, and have only ourselves to blame.

Dan: The quest for communication seems to be pretty much mutual on both sides. We both depend on each other, I guess. At least I would like to think so.

X: Their power is terrible and absolute. And yet they are tender and merciful beings, and only [a few] of them ever contact this crude world we live in. We have a long road ahead of us in order to change. We have a lot of changes. We have to do it, as they have understood with us. We have to do it in our way with their guidance. You have said it very correctly: they cannot do it for us.

Dan: What would be our purpose, then?

X: No purpose, [or perhaps] one that's clear in the Bible. I think Lucifer wanted to make man obedient through power and force, and yet the forces of light gave man freedom of choice. So, I know what it is to live in an atmosphere of iron rule—orders are orders, and there is no room for compassion. You obey, carry out your mission, or you are the enemy yourself.

And that was the iron rule of Adolph Hitler's regime. So in that sense I experienced the same comparison—they are the forces of darkness—absolute iron rule, enforced order. On the other hand, you have a sense of responsibility and freedom of choice. That is the difference between the powers of darkness and the powers of light. We have to help ourselves first before we can be helped.

Dan: I've heard a lot of things about some of the inner inhabitants being from Atlantis. I understand through other sources that quite a few of them before the deluge went into the interior. And my other question is, did we originally come from the interior and migrate into this lesser world?

X: This is actually the fourth and the last world. Those in control of the inner realms now are going to re-seed the planet and make it beautiful as it once was. Because all of this inundation, fire, and destruction took place before there were certain races that were allowed to enter caverns [that were previously] sealed, and they subsisted on a type of energy that the Arrianni possessed. You don't need food, but you are in sort of a catatonic state, unaware and not feeling the need for food and necessities and so forth. And after the cataclysm was over, the energy source was retracted and they were allowed to return to the surface world, as an example. [Their message is] "You see what you have done. Now you have to go out there and rebuild it. Have you not learned anything?"

This is the fourth [world] and the last time that's coming now. Many native peoples have legends. I'm sure you are aware of this about being in caverns and coming to the surface world. This is after they were placed there in a catatonic state. They went out and found a world wrecked with chaos and havoc and yet when they returned, and I'm speaking in time [...]

You see, time to the Arrianni is not relative as it is to us because they can go on and on and on. They can perpetuate their life for as long as they wish it. These various races that were gathered together and preserved were then cast out into a world of dark, harsh conditions. And when they walked out again into the sunlight and found the beginning of a new world, the trees were again growing—nature. There were flowers again. Animals were released. Sources of food supply were there—limited—but not in [such great] quantities that they did not have to make an effort to obtain it.

This is part of the cosmic plan. For people to understand it they must learn to live in peace. But the mercy of the cosmic forces gave them sustenance when they were released from these places. They didn't go there and just live in a cavern of some type. They were in a catatonic state, a sleeplike state, and a kind of dream world where they were conscious yet not feeling. Do you understand what I am saying? When the time came, the openings were reopened and they were allowed to exist into the world again. Once this planet had vast populations, far, far more than exist today, but in a totally abundant environment. You had only to pluck from trees and gardens everything you wanted. Nothing cost. Everyone was his or her brother's keeper. Everyone cared. There was no money, the rate of exchange, all of these things, supermarkets, and all the things we have now. And also there were no weapons, mankind never even thought of war until this catastrophe occurred, brought on by the forces of darkness. [Now] we have been engaged in combat for a long time. So those who come out of it will be the mettle of tomorrow's Garden of Eden. That's about all I can say about that.

Dan: Very well said, too. What proof do you have that they exist? Well, you've talked to them and had your own personal experience. What do you call these people? How shall I refer to them—as beings, creatures, or just people?

X: Arrianni is their correct name.

Dan: Do they live here?

X: They were on Earth in limited amounts at one time. You say that you admire Icelanders and Germanic peoples. They were descendants of the Arrianni. They had a fantastic [amount of] knowledge and it [is borne out] today in the German race of people, the brilliance of what they do. You can name many things and it has been created in Germany. Most of all the great inventions, all great strides forward—what we call strides forward—were made by the Germans, the Nordics, the Swedes, who are all descendants of Arrianni stock but diluted by a type of race that was here a long time ago.

This is what the Führer always said in his propaganda machine: "We are the master races. We are the Aryans, the only race on Earth that is capable of doing great things." That was all so much for the young, the youth. I even believed it once myself, you know. These Nordics [migrated] in the old days [to] what we now call the land of India, which is all vastly changed.

The Aryans came from the north aperture down, migrations of them, as teachers, and the Sanskrit speaks of them. And the word itself, Aryan, has survived. It's probably one of the oldest words in the world that is unchanged and in any language it is the same. That's why Germany has the strength of scientific genius running through it. At one time, such people bore the genetic code of the gods. It is not that they are racist in nature, but this is why the Order thinks about keeping the genetic code intact. So that it isn't further and further diluted to where it becomes meaningless.

Dan: Can the Arrianni be contacted?

X: Yes, the crystal contacts, if you truly believe it, [are found in] a remote spot. You will fail three or four times but eventually make it—the contact. It will be like a voice inside your mind speaking to you. They will answer your questions, and [from] those that don't regard [you] as worthy, you will receive nothing but silence. These beings are only concerned with highest metaphysical points of view they think are constructive. But if you are destructive in nature and think destructive thoughts, you will receive nothing.

This is what I said before. What [confidential name] has said [is that] those who have genetic racial memory are recognized, and are drawn into this immediately; others will find no interest. And we know that the number will be small. Because of the small number, there is great strength. It is like unlocking the atom unsuspectingly. This tremendous power is there as a mass of material [and] remains [so] unless there is a way to unlock its true nature. This is what we intend to do. We will gather those who are truly interested in becoming like a godlike race of men again.

If we can only establish [ourselves] and become worthy, we will receive help from the Arrianni and in the end the great gift. And that gift, you know, is the Garden of Earth. Where there is no war, no suffering, no pain, no sickness, and all is beauty and joy. That's the attainment. They can offer it. The Arrianni have it already inside of the earth, the Paradise that exists there. But it is not something to be given.

Referring to the Bible again—that chaotic old record—it says in there that you do not cast pearls before swine. Because what would they do with them? Tromp them, urinate, and defecate. And this is what most human beings—oh, I hate to say this—most human beings on Earth today are a type of swine. They just—with only one thought in mind they run to the feed [from the] trough of materialism and stuff their bellies and bloat.

Dan: What scientific experiments do you know of that demonstrate that the Earth is hollow? I am thinking of some lab type of experiments.

X: Those can be obtained if you write the Institute of Geology at the University of Moscow and ask about the MOHO project. They have drilled into the Earth something like ninety miles. And they have found biological life, minerals like they have never seen before. And they don't understand how this can be. It can only be that they are encountering segments, pockets of the Earth. Not the inner realm itself—that would never be allowed. But now they have found [a temperature aberration]. They always thought the [deeper they drill, the] hotter it gets, but they [have now discovered] a point where it begins to cool radically. You can write them.

Dan: How shall I request it in particular?

X: Just request this information and say that you are a scientific writer and investigator. Say that you have heard from the scientific journals about this tremendous experiment of which a lot is highly classified, and that you would appreciate any information that they can give you. They are very cooperative. The science in the Soviet Union is a free science. It is even greater than what we have. The scientists in the Soviet Union are regarded as very high, prestigious persons. They are given everything. They live like little gods.

When World War II ended, the Soviets were brutal to common German soldiers, but scientists were handled with the greatest kindness. They were taken away to the Soviet Union and a lot of the [greatly advanced] technology that the Soviets have today is from German minds. The science that they have is even greater and freer. They have vast resources. There are no limitations on ideas, and [scientists] are not exploited as they are in this country, but encouraged to find the truth. That is the most important thing, to find the truth. Not how to make something to make money with it. But what have we in this experiment here? What does it mean? What can it mean? How can it be beneficial?

Dan: Any particular department? Any person you know?

X: Just the Institute of Geology, the University of Moscow, the MOHO projects. There is this vast drilling. They found all these strange biological life forms from core samples. They are so fantastic in nature. They were not even aware that these things existed.

Dan: Yes, that sounds interesting. I would like to travel there someday. Some of these questions are my daughter's questions. She is a very special person. Her name is Zephera; I gave her this name so that I would never ever forget the ancient teaching that has been around me for such a long time. Her name comes from an ancient teaching that is familiar to us. It comes from the ancient Kabala.

In what way do you communicate with them? And how do they communicate with us?

X: Most of the communication comes [from] Ritters who travel the world and [have] special means of communications. This is touching on a little bit of classified information, but the contact is there.

Dan: I guess if anyone needs to know, they will make it known.

X: They will only suggest their noble ideas.

Dan: Here is another question from my daughter. The reason I refer to my daughter is because she ordered a book on UFOs from school. She was reading through it and she brought this to my attention. [I showed the book and turned to the bookmarked page.] Her question was about the men in black.

X: MIBs. Yes, they exist. These people are diabolical agents who work for governments, right on this Earth. They use special weaponry, drugs, and so forth to confuse and confound those they consider their enemies. They are not a good entity.

Dan: This source says that the MIBs are the ones connected to the Hollow Earth and that they are the Aryans. This doesn't seem to be a likely cause for enlightened ones.

X: Dark forces can be so clever. The greatest weapon that they have is to make you think that they don't exist. That is nonsense.

Dan: Do the MIB's really exist?

X: They do work for governments in this capacity. Nothing good can come from them. Their rulers are agencies like the CIA and others of this nature. They have vast power. I hope you never encounter them. Because if you become their target, the strangest things [you] could imagine would happen. For instance, say you were living in an apartment, and you did something that was threatening to them. You would become nonexistent. [You would] disappear, [your] friends would not know what happened to you, and very likely the apartment would be changed in some radical way. Possibly strange occupants would be there.

Your friends would come to the door and ask for you and the [strange occupants] will tell you, "I do not know what you are talking about. Are you a madman? We've been living here for the last ten years. This can't be my friend who lived here. No. Now go away before I call the authorities." And all the neighbors will tell you the same thing. "I never heard of this person." You would simply disappear and as if you never were there. And God knows what happens to you. That's what these men are like.

Dan: Interesting. Well, at this particular point you would have to wear the full suit of God's armor—something I wouldn't [find in] my own closet. I know my daughter would be interested in that when she's ready. She is a resilient young woman with an amazingly intuitive mind. Now I refer back to the Arrianni. What kind of government do they have?

X: They are governed by a genetically engineered race of biological robots. They have a set pattern of control and there are no infractions, no political connections or manipulations. The order of life is complete and they have eliminated war.

Dan: So they have created a being or a central power that maintains a complete order?

X: Right.

Dan: So a being or a central power certainly suggests—

X: Complete order.

Dan: Yes.

X: Yes, you have that right and very well.

Dan: So they become that law within themselves, and so the possibilities seem to be unlimited. Please describe what the Arrianni look like.

X: Yes. The Arrianni are tall, blonde, and beautiful. But the biological robots are a type of humanoid. They are servants and are completely obedient and faithful like dogs. They do every task willingly, cheerfully. It is like a good dog—they mind you and love you. And want to be with you.

Dan: Yes, the dog is the only creature that loves someone more than itself. Are they vegetarians?

X: Food from the sea is quite welcomed. Flesh otherwise is a forbidden thing. I understand they have animals there in the park, [in] areas that we wouldn't even begin to be able to describe. [Areas of] immense beauty filled with magnificent flora and fauna and animals that are there purely for the enjoyment of their beauty and so on. No one is looking them over and saying, "This is a large steer. It will be growing soon, and then I can sell it for X number of hundreds of dollars to the meat packers. They never even think this way. Animals are objects of beauty. I understand there are llamas there and all types of strange birds that we no longer have in this world, the outside surface.

They condone only those of us who have the misfortune to be in the circumstances we are in here. But staying with a diet of vegetation and seafood is sufficient.

Dan: I guess they have no need of fasting or cleansing?

X: No. Their whole world is filled with this vibrant magnetic energy that is so little understood. They have the physical makeup of gods. They have no need for cleansing or purifying. There is no more disease. No diet. These foods are beyond our understanding. These are the gifts they have to give us if we can only become worthy. Once in an age, they send a master from the inner realms here. The master can be visible or invisible. Because they have devices, I've been told, that change the refraction of light rays and the magnetism that surrounds us. When they cannot be seen, they become a tool of inspiration, like someone who has a pure and nobler approach than mere men have.

In our own time, there was one man and that was Albert Einstein. You see, he received instructions from one of the masters. That's why he is so great and so little understood in his work. He was told how to do it. You see, men and women just don't sit down and say, "I am going to create a new type of energy," and so on, because we want to do it. It is inspired and it's received and comes from another source. It is given to them. It is given free. Well, before Dr. Einstein died, he said, "In view of what they have done with my work here, I would rather have been a common plumber, than to see what they have done."

So then I remember, shortly before [Einstein] died, he was in Trenton, New Jersey, and asked about his regrets over the development of the atomic bombs. "Oh, yes, I am very regretful," [he said]. And they asked him what the next war would be fought with. He said, very philosophically, "I can tell you what the next war will be fought with and the war after that one also." And they said, "Well, then, what will it be fought with, this one and the one after that?" "If the war comes, which I hope to God it never will, we will use nuclear weapons in the third, and the fourth world war will be fought with sticks and stones." He was an inspired man, but his knowledge—his gift to the world—has been misused. Always this happens for some reason because of the forces and the powers of darkness. What God has set into motion, they intend to make static and destroy. This is the part again where they [of] the inner realms will help the few if they can become the worthy.

Dan: Do the Arrianni contact humans?

X: Yes, they do. My own experiences [show this to be] true, and [there are] many others, and they always happen in a remote place. There are no witnesses of any kind, because they are not here to be sensational. But they are here to give information and also knowledge about certain beings. They give half of the information and the other half you must rationalize yourself. There again, there is self-help. The effort you must make.

I referred to Dr. Einstein, who was given great inspiration and told [...] and yet he had to spend hundreds upon hundreds of hours with mathematical equations, all handwritten. Now, a computer can do this, but he had to do it all by hand on a blackboard, searching for hundreds and hundreds of hours in a little room. The man evidently was looked upon as an old eccentric. This strange old Jew who was dabbling in things that [others] would consider meaningless, yet he changed the whole world. And he was just one man. It is always the small group that makes all the changes because they have become chosen. And they get inspiration and knowledge from the beings from the inner realms.

The Arrianni have these things and [they] are evident in every way. They are not smart or proud by asserting what they know, or that we know nothing. It is not that way. They are our benefactors. They have a certain compassion for us, but they will not become involved with our violence, because they have overcome war. That is why they live in a closed environment.

Dan: How do you choose sides? Okay, you mentioned this before, is it possible to fly into the inner realms?

X: No longer. A magnetic maelstrom will destroy any attempt [at entry and] will destroy the sturdiest craft we can ever imagine. All of those entrances are sealed. It is a gate, a barrier, and it's there as an absolute force of magnetic energy.

Dan: This is an invisible type of force?

X: Byrd encountered this with his old C-47 aircraft when he flew. The first warning of this was [that] the gyroscope began to wobble and become erratic, although it had electric power to stabilize it. I'm sure you know how they work. The instrumentation went wrong and the vibration began to affect the aircraft and then, when he thought something terrible had happened, the aircraft was stabilized in the air and began a downward decent. The craft was opened and alongside came Flugelrads telling him they were taking over command of the ship, and he was guided in.

And that was in the secret diary that Byrd had written and kept secret for so many years. And he was looked upon as a lunatic. But it happened. And he was chosen and was actually allowed a few moments at the time with the master of the inner world, telling him to go back and tell the world not to try to develop this atomic energy further. "It is not for you." It is the same parallel as telling a child [about] a stove. "It is very hot. It is hot, and don't tamper with it." That's what he was told in so many words. And what happens to the child that ignores? It's inevitable. This is what Byrd was informed, and [he] was pounced upon by the Pentagon military men. And he was discredited.

The information, all of his logbooks, his films, everything he had made was confiscated and probably destroyed. But the secret diary that he wrote escaped it all, and in reading his pages, something like the truth must and will come out. What it means to man is that he must determine the truth himself. [Byrd wrote] of a twilight darkness coming on. Evidently he had a premonition of death and he wrote [...]. He kept silent, and remained silent all these years. But now he must release this information, regardless of the consequences.

Dan: Have you read this diary?

X: Yes, indeed. It exists and it is in a very safe place. There have been copies, of course, for those who wished it. The original diary—I think there are forces that would give very much to place their hands on it, to destroy it. Of course, even if that is not so, they can always say that these are the writings of a demented, sick old man who had hallucinations. He was crazy. Do you want to believe that? Or would you say you just don't know? But now, what is this really? And how much is it really the truth, if it were for you to determine?

Dan: Does the Admiral's family know anything about his diary?

X: It is my understanding that they know nothing about this. He kept this totally to himself in agreement, evidently, with the powers and the authorities that obviously threatened him. They told him, "You are a military man. You are bound by an oath to keep this matter secret. If you don't, many [dangerous] things will occur." So I think the old Admiral had a love of life [and] respected their wishes until he was certain that they wouldn't touch him. He knew he was going to die. Then he released this document to someone.

Dan: He personally released it?

X: I don't know. But the information did come out.

Dan: Is that document here in the United States?

X: We had at one time a limited supply of them that we simply mimeographed and sent out to those who wanted them. And of course, since that time, we have been warned not to tamper with them further. So these documents are no longer around, except those [copies. Those] who are fortunate enough to have one can make copies of the copies.

Dan: May I read a copy?

X: I shall see that you get a copy of one. Now, what else do you have there?

Dan: You have talked about different countries. Have different countries flown into the inner realm? Since they are sealed now, I guess nobody can get in there.

X: No one. But several principal nations are involved. Watch stations, commonly known as weather stations, measuring meteorology, temperature, and seismic activity, all of these things in disguise. Their main function, though, is to observe comings and goings of strange craft. They keep an accurate log. Oddly enough, you say, you are an air force veteran. They had an interesting project, Blue Book, which was suddenly dismantled. It was an investigation of flight activity of strange craft. You remember it?

Dan: Yes. It comes to mind. Do you think it would be practical if my research partner wrote to John Glenn in order to understand what information may be released? [President] Carter mentioned at one time that he would release certain information.

X: They would not because the president has very little power in this country. We have a form of government that the people believe functions a certain way. But it does not. The real rulers of America are those who control the oil and all these things. These are the real rulers. The president is only a figurehead, I am sorry to say. Certain of them are aware of this phenomenon that exists, but they do not quite understand. Therefore, they have feared these craft and what they mean.

Dan: How do the Arrianni relate to us? You mentioned they could land in remote places for selected individuals.

X: Oftentimes they select certain individuals because of the magnetic scan they can tell about people. If they were to land [their craft] publicly, the police, the media, and the army would fire upon them. They would try to destroy it rather than [be] reasonable. I believe, and many believe these forces, for instance, were fired on by stupid fighter planes. It was like a gnat trying to attack a fly swatter. [Fighter planes] have no power up there. There is no way they can disable one of [the Arrianni] craft. They are protected by a magnetic force that vaporizes any attempt of incoming missile fired. It would mean nothing. It would hit this magnetic force field and disintegrate. Firing, let's say, a 50-caliber machine gun—the projectiles would simply evaporate. Mankind does not want anything here from another world because they fear another kind of intelligence would take control of our world. That's what man thinks. Man fears he's going to lose everything, then attacks and destroys. This is the brutality of humankind.

Dan: That's been the age-old battle here. Many questions here I'm going to skip. I already have an understanding of them. Are the inner inhabitants friendly toward us? You've already talked about that. Sometimes I feel this attitude is in part of my responsibility, and I'm not listening. It's like when your wife wants to talk to you and you're pushing her off into the next room or you're too busy, not thinking, not concentrating or something. You're not being open, so perhaps she will throw something at you, slam the door, or walk out of the room.

X: To get your attention.

Dan: Yes.

X: Yes. You have a graphic example of what you say—to get your attention. "Would you listen to me? Would you stop what you are doing and listen?" Maybe she has to throw a small vase at you or something and then get your attention.

Dan: And maybe we can talk.

X: Yes. I wish this term UFO would vanish. I know it means an unidentified flying object. The true name is Flugelrad. That is the name of the craft. It's what they are called.

Dan: What does it mean?

X: It simply means a winged wheel.

Dan: That gives it some identity.

X: It is the identity of the craft itself.

Dan: Do the Flugelrads have a coordinated flight pattern?

X: The magnetic grids, as I say, have the power to deviate with their own energy. They are capable of flying to any geographic location. They are not like on a track. They have free flight. But they gather their energy from grids lying on the Earth—invisible magnetic fields. If man had that secret power, we would be in a lot of trouble.

Dan: I often wonder about some of these wars, like the one that took place briefly with England and Argentina. I don't know if it was a war of economics or pride or what the battle was all about. A lot of men spilled their blood, although for what reason—it still baffles me.

X: The change comes, yet things remain unchanged. And in the meantime a lot of people, as you have said, have spilled their blood and died for something totally meaningless. This is what we have to change. And the change can only come through great invincible power, which the Arrianni possess. This is what would have to be done in that respect.

A select group of totally dedicated spiritual men [are needed] who can at last sit down in the inner realms with the master, the Prince himself, and say, "We are here to try to bring back what we once had. [Although] we are only men, our desires and hearts are pure enough to sacrifice ourselves for whatever is needed for the whole of humanity, which is worth saving."

Once that could be done, the Arrianni could send the Flugelrads into the air and any opposition would be utterly destroyed, and whoever is left would not be treated as slaves on a slave/master basis. It would be a matter of assimilation of knowledge that would be given to the human race and a restoration of certain magnetic factors—the creation of the garden again, and all would be peace.

To put it in other terms and philosophies, some religious groups in the world say that someday a Messiah will come back. The Messiah will have absolute power and anyone who defies him will be broken with a rod of iron. Well, they have the right idea. They have the right idea.

Dan: Yes. Each religion does present some vision and some concept of restoration, bringing us back to our original state.

X: We are talking about the Bible again. In Ezekiel, it speaks about the wheels within the wheels, and the round ports and the faces of men. Those were Flugelrads that the prophet saw, but he didn't recognize what he was looking at. Say, you yourself, [what] if it were possible to place you in a state of suspended animation for a thousand years? And the next morning, one thousand years had gone by and it was like a night of rest for you. And you opened your eyes and went out of your building, or whatever container or capsule you were in, and you saw the world and you saw such fantastic things.

How could you even begin to describe it? This is what he saw. He encountered something. He was part of an archaic culture and was confronted for a brief moment with devices that were hundreds of centuries ahead of him. And this was the only way he could explain what he saw.

Dan: Are there other races besides the blonde race?

X: No, only these biologically-engineered entities look different. For instance, some of them are brown and have almond-shaped eyes and black shiny hair, as we know them. And there are others of different stature and different shape. These beings are keepers of vast generating sources of energy. These beings have been engineered to be involved in certain force fields of energy that will not harm them because they are engineered to withstand this influx of this energy.

You have physical differences and changes, but basically it is the design that is the difference. In other words, humankind has now developed atomic energy-generating stations, and those that work in there have to have special clothing, protective devices, and even mechanical hands to handle various types of radioactive material, you know that. Well, you see, the Arrianni are so advanced they don't go this route of vast mechanization and special equipment. They say, "This magnetic energy that we have is a very dangerous thing."

The Arrianni have genetically engineered biological beings that are able to withstand this and to work with it and serve—that is their purpose. And you find these [other beings] in various places, I understand, in their realm. You can call them a different kind of race, if you like, but they are engineered—genetically, biologically—to work with very dangerous energies without any harm to themselves. They are working with elements that they themselves are a part of. Therefore, they are immune. Merciful, isn't it, and beyond brilliance. To create something to serve out of the very object that is being served. All matter and energy has intelligence. We know that. We know very little of it, but we do know that, don't we? We are advanced that far.

Energy is intelligence in itself. An accumulator—it's called a battery. You can discharge it to where it no longer seems to have the slightest impulse. Then that energy will regroup and come to its own source and kind—if you let it be for a few hours. It is not much, but then we turn on the ignition and we do get a response. How does this energy know how to return and to regroup to its own kind without any intelligence? We call it a physical law. It is innate intelligence of matter and energy itself. These beings are made of the same principle.

Dan: Captain Bruce Cathie, in his book, Harmonics 33, mentions that at least three Flugelrads have crashed in New Mexico and Arizona. Sixteen bodies were removed. They looked like humans except were only thirty-six to forty-two inches in height. This information was given to a journalist named Frank Scully by a top scientist who worked with the crafts.

X: Interesting, you note. They looked like humans. They are genetically engineered humanoids, robotoids, however you wish to classify them. They are not human beings. They were pilots and crew. Their craft was a very small auxiliary observation craft, and if it's true then it is only possible that some malfunction must have occurred of unknown origin.

Dan: I've heard a lot about the mysteries of Mount Shasta. It is a beautiful mountain I love to visit once in a while.

X: There is a magnetic power in that place and area. It is one of an energy-gathering source. There are people living inside that mountain. Not people, but beings. Did you know that?

Dan: I've encountered some experiences on that mountain myself—incredible things. I've yet to return there and spend more time.

X: Anyway. They are there for a purpose. Those who are there, as I understand, are classified as teachers. They are a race of masters and they may come forth someday, but the world is not ready for them. I can say very little about them.

Dan: Captain Cathie also told of a US sub that was taking pictures of the sea bottom, fifteen thousand feet below the surface. They photographed an aerial-like object that was sticking up from the otherwise featureless seabed. It was entirely symmetrical, two inches in height, with six crossbars located at 59°08′ south latitude, and 105° west longitude. How close are these coordinates to the opening to the center you mentioned?

X: I think what you are speaking of are monitoring devices. The Soviets have developed a high technology and they seen these disk-like objects. They are like sonar in a way. Any U-boat passing in the area—the turbulence of the propellers and also the noise in the internal machinery—is monitored by these and sent back through a relay system to certain bases so that [...]. The United States has the same. They monitor the coming and going of undersea craft that way. I don't think there are any other consequences.

Dan: You already mentioned the grid systems. Cathie says this is supposedly a network of magnetic antennas implanted throughout the Earth in very precise locations. UFOs are said to operate in harmony with these grid frequencies. Cathie says that UFOs have been repairing grids that have been damaged since the fifties. I'm wondering if there are all kinds of grids.

X: Yes, there are systems of grids called [...] as a beacon. These grids are fully functional magnetic automatic guidance system for rapid deployment of Flugelrads from one point to another without any control. The only control necessary for the Flugelrads themselves, as I understand, is in free flight or to the outer atmosphere. Then there is a definite control. That's about all I can tell you about that. It is a monitoring beacon system much as we would use lights for navigation from the air. It is something on that order.

Otherwise, I couldn't comment further on it. I know, in a few cases, that power lines, for instance, have been used for refurbishment of magnetic energy, and evidently, in emergency conditions when it is necessary to obtain a tremendous amount of energy quickly. This has only been in a very few cases of very small observation craft. The large Flugelrads are completely self-sufficient in every way.

Dan: Are they independent of the Earth's outer surface?

X: Yes, they have an unlimited flight.

Dan: They are not totally dependent on the grid system, then.

X: No. They can penetrate the outer atmospheres, and I can't even begin to imagine where they go. This is something that is not known.

Dan: The book, The Hollow Earth, by Bernard, mentions a small, brown-skinned race living in the interior that may have been ancestors to the Eskimos. What kind of information do you know?

X: I can tell you exactly. A very long time ago, those that we call Eskimos, they were brought out of certain parts of the interior after the cataclysm. They were brought there in a Flugelrad and left some time later. These are descendants of biologically engineered beings that have adapted to external environments and circumstances. But they did at one time originate in another place.

Look at them from a physiological viewpoint—doctors can tell you that there are many differences. A physician, one who studies physiology, will tell you. There are minute variations, but it doesn't matter. They are just a little different than we are. But it's because of their early beginnings that they are different. You see, they can adapt to the surface world with minor changes.

Dan: So they adapted themselves in a remote part of the world?

X: Yes, where it is cold and bitter, [and where they developed] the ability to withstand temperatures that most people cannot, to live on a diet that most people cannot. And they have within their mental makeup a totally different outlook. That's why you find none of them ever coming to intermingle with the rest of the world. They want no part of it, because they remember, from a very long time ago, something that happened that was very terrible. That's why they remain isolated for the most part and don't want to become any of this. It's their nature.

There was a settlement some time ago in what is now called Canada where a group of them lived, and they were concerned about the encroachment of civilization, of what they deemed to be unjust laws [that] threatened their existence. Among that particular group, they still had what we would call a holy man, who had the power to communicate with the inner realms. They send for the Flugelrads to come and take them away. And the request was granted. Did you know about that? Yes, it's on record as north, some village. Totally disappeared, every man, woman, and child. The Mounted Police [R.C.M.P.] found a totally abandoned Eskimo settlement. Not a trace of one man, woman, or child. Food was still left where it was, and the sled dogs left tethered. They had left with no explanation. That's what happened.

So, in isolated instances, certain things happen. Probably the same happened to the fliers, the American navy fliers in Florida several years ago. A whole squadron disappeared. You know about that, I'm sure, and of the strange disappearances for reasons that we cannot—we can't question.

Dan: Is this what is referred to as the Bermuda Triangle?

X: Yes. The best I can explain it to you is that on the bottom of the sea, covered over by a mountain now, was an ancient magnetic generator from times gone by. I don't know how many hundreds, many thousands upon thousands of years. But it is still functioning and at times it becomes erratic in nature, causing strange magnetic disturbances in that area.

But that is the source of that great mystery. It was a machine of the ancient Atlantian, if you like. [It was] one of the devices they possessed for moving objects, transportation, and many things. You would think going back that far in history that this was a race—a chaotic old race, but they were more advanced by far than we are, at our point here. We have survived this last great cataclysm, [yet] we don't remember many things about it because it is in our nature to suppress and try to forget what is unpleasant. But that is the mystery of the triangle.

A great crystal remains as a magnetic generator on the ocean floor, which erratically functions from time to time because it is no more a control factor, and therefore these disturbances occur.

Dan: Yes, that's interesting. My oldest brother, Fred, has been researching such a crystal, too, and claims it's from his Atlantian memory. He's in and out of this thing in a confused way and sometimes in a profound way. He's desperately trying to balance this knowledge in his life and the memory that haunts him.

X: Everyone must try to satisfy their own needs when they try to rationalize. That's progress.

Dan: Are the Arrianni undertaking any construction on the ocean floor or on the ground surfaces for Flugelrad bases?

X: There are bases, sure. They come from certain entrance- and exitways. If you would, call it a base, but it is like having a secret airfield. They have no need of an airfield as we know it. Their technology suppresses anything like that. That's why the Russians, Americans, English, and so forth […]. There are certain regions in Antarctica watching these comings and goings. They simply have no need for such.

Dan: What about the far North? I have been drawn to Iceland like a magnet. I've had many dreams and psychic experiences about the origin of Iceland. The country and its people are all familiar to me. The legend of the Norse Gods seems alive and rich in the hearts of its people. I have visited the country several times. I understand what the author Kurt Vonnegut meant about Iceland. It was his idea of Utopia with a strong sense of family. I have made great friends and met some very interesting and well-known people. They are a beautiful, peace-loving, a mystical race of people rich with the memory of their origins. I could write a book about it.

X: Yes, of course. I understand.

Dan: Do you know anything about the mystical mountain Snaefellsjokull? I was introduced to Halldor Laxness. He was a Nobel Prize winner in literature and wrote a book called Christianity at Glacier. It was about a mysterious race of people that live under a glacier. It also is the mountain that Jules Verne made famous in his story, A Journey to the Center of the Earth.

X: About the only thing I could suggest is [that] you go there and try to find out all you can about it.

Dan: Before I went, a psychic told me—I didn't even ask her—she just started speaking to me. "You have traveled north once, and you've been traveling north in ancient times." And I asked her, "Well, how far north have I traveled?"

X: The Nordic Lore and the lure of the Nordic—is genetic racial memory.

Dan: Interesting. I do remember things—in ways that are not conventional thinking. Even my son had experiences about Iceland and dreams about being inside a pyramid.

X: You must know, then, that the shape of a pyramid is a type of generator, more correctly an accumulator of energy. It is very primitive but nevertheless an energy accumulator. Science, geologists have always thought that it was the extreme climate of Asia that has preserved the remains of pharaohs. But more important, an unknown for a long time, was the fact the shape of the pyramid and its design brought and accumulated a type of energy that's all around us, every place, in this room, everywhere.

There's an Austrian. His name is [confidential name] and he had expounded upon this very energy and its agents and [called] it argon energy. You could set it to a small enclosure of the correct design and receive very helpful benefits, and rejuvenation of tissue and body. He was arrested. He came from Austria in the latter forties, early fifties. He began to print understandable books. And he began to gather a large following of medical students. And he pronounced the idea of argon energy and was arrested by the American authorities, the Food and Drug Administration, and was rushed away to a federal prison. Shortly before he was due to be released, they murdered him and called it a heart attack because of the knowledge he had. All of his books, his writings, have all been totally destroyed.

Dan: My God! Someone is working awfully hard to destroy those who can advance the human race four-fold. This sounds like a great research to get into.

I understand there are six places in America and Canada that have gravitational anomalies. Two of the most famous are at Santa Cruz, California, and at the Oregon Vortex. Gravity and light waves are warped in these areas. Does this have anything to do with UFOs from the interior, or could these places be entrances to the center that are used now or were used in the past?

X: More correctly, it has to do with the mass destruction, the cataclysm that took place where full continents were displaced—elements fused, broken, transferred, and so on, account for these things. You see, in those times, a long time ago, we had the garden again. Everything was perfected. It was designed. The Arrianni are not a democratic entity. They are closed. The only apparitions that exist are those biologically engineered for purposes of work and handling certain elements that are necessary. No more than what we do. We use horses to plow our fields. And yet we don't mistreat them. But we do understand we are two different things. It's very simple to understand.

Dan: I would like to know more about the genetic racial memory and how the different races of mankind relate to this genetic code. To me the best way that I can understand this is that perhaps it's like a rainbow. A rainbow that exists with colors but the colors is not in conflict with each other. They are in harmony but remains to each to its own kind.

X: They're all to their own kind of vibration. The purple, the yellow, the red, the blue hues, each is different. And you thought of it very well. Yes, they co-exist side-by-side in harmony, each separate. You have it correct. Every race, every racial entity should be proud of what it is. Blacks should be proud that a divine creator made them black. They should not want to be white. Nor should any white want to be black. Or any yellow want to be white, and so on and so on. Everything here is working as it should. But look at what man has done by making all these alterations and changes. And all of the troubles he's run into because he would not let well enough alone. Isn't that understood? Hmmm. This is correct. Don't alter divine creations. The Heilege Ritters will kill no animal and they respect nature in all its aspects, and that nature includes man—very much a part of it.

Man has run into trouble because he is on the outside of nature instead of being in harmony with it. Therefore he has great conflict. He must remember one thing. He is part of nature also and not above nature. You often hear the idea: conquer nature or the conquest of nature. The most foolish mistake that any man or any race could ever make is to attempt a conquest of divine creation. In the time of the garden, all things were free and beautiful and wonderful. There was no need for conflict. Conflict only came after things changed. And so it is with nature today. The harmony can still be there, but we have to earn our way back, not our way forward, but a way back. This is what we have done. We have lost our way.

What is being advocated is not racial discrimination, but placing all things in their proper perspective as creation ordained it in the beginning and all was harmony, peace, and all things were plentiful. The animals walked amongst themselves as well as [amongst] man without conflict. Each was proud of its own beauty.

But now there's conflict. Men are not happy with what they are. They want to be something else. And this leads not only to total chaos, it eventually leads to destruction because of the constant conflict. It's very easy to understand that it is weak people who do become something other than what they were intended [to be]. And at the end it leads to grief, misery, and pain. So it is better to be proud of what God has ordained for you from each seed and flower of its kind down the line. Roses beget roses, lilies beget lilies, pine trees beget pine trees, and lions and tigers beget lions and tigers. But the one divine creation that was given the free will, that is the one that has gone wrong. Therefore, it is not the road forward but the road back to the garden because every movement forward begets more and more friction. That's in our own time—I've seen it well. From the 1930s the world was recovering from a great struggle.

But previous to the flood of Noah there was peace in the world with beauty and harmony. Then much later in our recent history the discovery of the continent here, America, inhabited by a totally magnificent race that was known as the Indian. Then came the clash between two cultures, and the white man came in with guns and committed horrible acts of genocide and murder to obtain the land. He justified his thinking by saying, "I am strong enough to take it. You don't need it anyway, so we'll just eliminate you." This was [considered] progress forward. And so it has been, from the bow and arrow to the cap-and-ball pistol to the repeating rifle, to the machine pistol, to the cannon, bombs, and now to the missiles. This is [called] progress yet it all leads to destruction. The road is not that way. The road is the way back. That's where we must go. It is back to the garden of peace.

Dan: This reminds me of a whole sequence of events that took place when I got married, a landslide of cosmic activity for several months, which was an eternity to me. For a moment, my wife was a reflection from a time that we both remembered. And another time we lived in the Golden Age, and we remembered that too.

X: Racial memory.

Dan: It surfaced. And I realized that we could not live in this world, and we just came tumbling right back out of that eternal state. It was all very painful.

X: Sometimes the realization of what you just said leads people to the ultimate brink of suicide because they can no longer face life. They have a flashback of when things were beautiful and wonderful, and they look forward or ahead and they see the cataclysm—horrible things. And they kill themselves. Do you realize there is a large number of suicides [for which] physicians here [cannot or will not understand] the reason and motivation? This is becoming alarming because the road forward is dark. And one look back into the past, if you can truly see it for a moment, just as you've expressed yourself, [makes] you want to go back. You want to find a way back. Isn't that correct? You would like to live in that time again, eh—back in the Golden Age?

Dan: With the clamor of ancient memories, I attempted to step back out of many horrifying cosmic experiences. I'm not so sure I can go back. I don't know if it's really possible. However, I did experience a vast cosmic memory.

X: Yes, you can. There's a way. Are you familiar with any of Cranach's paintings?

Dan: How do you spell that?

X: C-R-A-N-A-C-H. Cranach was an artist of the middle ages. I wish sometime that you would write to the Louvre, the great art museum in Paris. Write and ask them if it is possible to obtain a painting in Paris—a paper color reproduction by Louis Cranach. It would cost you four or five francs. Ask them for a color reproduction about the size of Louis Cranach's [original]. It is called The Golden Age. [He] is a very famous artist of that period. He has other works. Have you ever seen his painting of Eve? She has a long, slim body with long hair and the apple in her hand.

Cranach one evening had a vision. He was actually able to look back in time and see time when it was the Golden Age. It was a splendid, marvelous painting. It ranks with Rembrandt and any of those masters of those times. It will show you how things were and with lots of vegetation. Men and women swimming in clear, crystal pools of water without any clothing and trees with strange and beautiful fruits growing. A maiden is sitting down by a tree petting a large cat of some type. All of these things were without shame and in a sense of beauty—a real sense of beauty and contentment where all was golden. Will [you] look at that and see what it does? It is a joy to have in one's home. Because of the motivation of Cranach himself, he was able to see the Golden Age as it once was. This original picture is worth several millions of dollars. And it is enclosed in a hydrogen-filled case to protect it from further deterioration. But that will give you an idea of the Golden Age that we must go back to.

Dan: I, too, had a similar vision. I was descending from a cosmic state of mind. While tumbling through the vision in my mind, I came in contact with each race of people. It was very confusing to me and yet the vision was interpreting its meaning to me. It was a visual deception. I was one-sighted within my third eye, then tumbling back into my objective sight and awareness. I was no longer observing from my inner sight [pointing at my forehead].

While falling from cosmic grace, I became conscious and noticed how divided the various races of people were. I began to ask these questions of myself. "What about the Chinese?" I thought. "They will disappear," I heard a distinct but small voice say. I was seeing through my third eye from here [pointing at my forehead] as they were all taken away! To where and how, I don't know. This is what I saw and heard in a cosmic sense and what had been interpreted to me. I'm thinking, "My God, what about all the other races of men—the black, brown, and white? What fate is waiting for them?" Something was very wrong and dramatically changing. You're the only person I've ever talked to about this because it's such an explosive subject. It's highly volatile.

X: The forces of darkness are always quick to wail and gnash their teeth when they are confronted with truth. It is something you must expect.

Dan: It was so unusual. I still don't understand it.

X: Do you realize what a privilege it is to be what you are? It is a divine gift that providence could have changed at any moment along your genetic line.

Dan: It was some kind of way of observing. I was standing in the past, yet I was in the present and in the future. It was like walking a straight line and seeing into three worlds at the same time. I stood as a circle with no circumference. The vision was a disharmony, a boiling pot leading to misery, pain, and chaos. It was total disarray on the face of the planet to such a degree that only an act of divine providence could restore order, peace, and harmony within humankind's heart and mind.

X: The time is coming when everything will be separated again. Those that remain racially pure will not have to endure the horrible trials that are ahead. They will simply be taken. The rest—I told you what would happen to them.

Dan: Do you mean the Second Death?

X: Those that remain pure and are proud of what they are, those of the yellow races, the red races, the black races, and the white Aryan race—those that remain pure will find their salvation in their own purity. It is hard to accept and understand, but nonetheless it is true. The truth is not always easy to assimilate.

Dan: It is an unfinished plan.

X: Yes, it is an unfinished plan. You are so correct. I have to be leaving here very shortly now. Take this address down. Ritter von [confidential name]. Tell him about our meeting here and something of the things we discussed. Tell him that he has been appointed to be your mentor in all questions pertaining to the function of the Order. And include my name and this is at the request of Ritter von X. So this I have done for you. I am going to have to leave now.

Chapter Nine

◈ What Does it All Mean? ◈

When the student is ready, the master will appear.
—Gautama Siddharta, 563-483 BC

I have re-read my notes of our conversation with Ritter von X from a decade ago. Nothing has changed since then. If anything, we are even more aware that a cataclysmic event of some kind must occur. After the destruction of Paradise Earth, man has been forever shipwrecked. I realize that only through man's memory is he called to the genetic racial memory, better known as the Akashic Record. The Earth is like a seriously damaged spaceship that must be brought in for repair. I am still pondering the lessons given me by Ritter von X on that brief weekend so long ago. What is the meaning and importance of the dilution of the genetic racial memory or the Akashic Record?

There was a knock on my door. I was almost finished going through all the printouts of the conversations. Still holding the documents I was reading, I continued aloud. I fumbled for the key and opened the door. *"The forces of darkness are always quick to wail and gnash their teeth when they are confronted with truth. It is something you must expect."*

I looked up to see my next-door neighbor, Wanda. She was smiling, dressed in a faded apron, hair in pin curls. She extended a freshly baked German chocolate cake toward me.

"Care for some 'daily bread?' Or is German chocolate cake considered one of the dark forces you just threatened me with?" She tried to be casual, but I know I sometimes scare her.

"Hear me, Sister!" Papers sailed through the air and landed all over the floor as I swept the cake dish out of her hands. "Truth!—German chocolate cake is definitely Truth, Sister!" I take on a German accent. "You can trust me on this, Wanda. This is fact, you see!"

"What on earth are you into now?"

"Not 'on earth,' Wanda, darling," I was still in the German mode. "In Earth. In Earth!" In my regular voice, I continued. "Just the last of the tapes Diane transcribed back in 1983. We're right down to the nitty-gritty now, Wanda."

"How's that?," she asked, pouring us both a cup of coffee. Then I remembered I had sworn off the black poison. She dumped my cup into the sink, refilling it with hot water from the kettle.

She rummaged through the clutter on my kitchen counter and finally procures a bag of Red Zinger from the cupboard. "You are down to what nitty-gritty, Dan?"

"How much of this Ritter von X stuff can I use? Most of it is so good. Like, look at this." I showed her some of the pages. We ate cake and drank our hot liquids as she flipped through the pages.

"Wait till you get to the part about the rainbow," I said. "The rainbow becomes a metaphor for the different races. In fact, it's more than a metaphor. It's a physical phenomenon. It's simple. The people of the Earth were destroyed. Noah gave a hundred and twenty years' warning about the catastrophe. People in those days lived a thousand years and more. Paradise lost! But not forgotten. We have been left to find our way back to Paradise. Humanity lives on the outside of the laws of nature. At one time, we lived within the fellowship of God and His Kingdom of Paradise. Jesus Christ reveals it in his prayer to God."

"You mean the Our Father—"

"Yes, 'Who art in Heaven, hallowed be Thy name, Thy Kingdom come, Thy will be done, on Earth as it is in Heaven,' et cetera. God's Kingdom on Earth is Paradise restored once again. Currently, our consciousness is measured by the outer circumference of the Earth's centrifugal force. We constantly retrace that circumference, marking our stay on the Earth in years."

Wanda stared in disbelief as I stood, my arms outstretched but bent downward sharply at the wrists. I begin a slow spin, moving my arms alternately up and down as I continue, hardly taking a breath between sentences. "We live in memory of each other. Humankind is like an ant walking on the top rim of a glass. It just goes around and around in circles in repetition, thus only advancing certain aspects of its being. Humanity is riding upon the outer circumference of the Earth's centrifugal force. So when consciousness is raised, gravity changes accordingly. There is a mathematical formula in this somewhere. Then we move into an inner realization. It's like standing on the outermost edge of a spinning wheel and realizing all the sensations of that force. It's a giant merry-go-round! We all perpetually go around, around, and up and down. The result is obvious." I finally took a breath, ending my dizzy imitation of a merry-go-round.

"Obvious." Wanda repeated. Her eyes were glazed as if in a trance.

I continued, hardly noticing. "Obvious. We let go and fly off the wheel and eventually decay into the illusion of death. Our years on this Earth are like concentric tree rings. Nature is here to remind us where we are in relation to its forces."

"Death is an illusion?" I was losing Wanda now, I could see.

But I was on a roll. "Through the Christ consciousness, the human species is kept in existence. Another example: look at an apple tree and contemplate the invisible intelligence that travels through the body of its organism. Notice how this godlike force moves through the whole form of the tree and into the first stage of what we call spring. Thus, the tree begins to blossom forth with leaves. Then it forms flowers. The flowers fade. Then early stages of fruit begin. The cycle proceeds. The fruit is ripe with the life-giving seeds from which it was created. Again the cycle proceeds. The fruit is gone and the leaves begin to fade. From spring, we traveled into summer, then into fall. All the leaves loosen and tumble from the life-giving force and fall back to the Earth and enter an even bigger cycle. Cycles within cycles are various degrees of consciousness."

"Would you like more tea, Dan?"

"Sure. Then the winter creeps in as a silent darkness when all life seems abandoned. The tree is stripped of all life and seems to be cut off from the life-giving force itself. But it's not! While the tree remains barren, the cycle begins to repeat itself again and again!"

"Yeah, so?"

"The Earth year is a repetitious cycle. If you raise the consciousness of that cycle you ascend another step toward immortality."

We stared at each other silently for a moment.

"Do you have a point here pretty soon, Dan?"

"Wanda! That is the point! Don't you get it?"

"Dan, you aren't telling me anything I don't already know. Isn't there a huge 'aha' in here somewhere?"

"Wanda," I sighed and took a drink of my Red Zinger. Then I began very slowly and very precisely, "We live. And we dwell. And we have our being on the outer edge of this force. To be in the center of it would mean to transcend our present-day reality. Okay?" I looked to her for confirmation.

"Yes, I'm with you."

"So, death would no longer exist. Yes, if we lived in the center of that consciousness rather than on the edge of it. Death would no longer exist, Wanda! That's the point! Even the night sky would disappear. All living things would be in a transitory state of ascending consciousness toward an eternal life existence. Every living thing on Earth would be subject to this cyclic change. In this world, we are just like the leaves. If you could observe from the center of this invisible intelligence, you would be seeing from the interior, outward. In the last chapter of the Bible, it reads, '...and on either side of the river, was the tree of life, which bore twelve manners of fruits, and yielded her fruit every month.'

"However, if we gravitated toward the center of the centrifugal force it would no longer have an effect upon us as we experience gravity now. So, a fruit tree growing in Paradise would produce fruit every month. The four parts or seasons of the Earth's revolution would have enclosed within the center of that force. It is like the hub of a spinning wheel." I paused.

"And what does all this have to do with the rainbow?" Wanda seemed anxious to get back to the first point.

"The rainbow is the symbol of the fall of humanity. Paradise imploded and collapsed as a result of the great flood—induced by our own efforts. Please understand this is a physical phenomenon that literally changed the surface of the Earth. A rainbow is a rainbow because each color shows itself as a separate vibration. Each color is a different frequency and assumes individual identity. The colors are the ingredients of the white light. Likewise, each race has its own frequency. This is what we must understand as the difference. The rainbow reveals its forgotten meaning."

"That's really important in a situation like this. You can't put yourself in the position of being on one side or the other in a race issue."

Wanda was only pointing out the obvious, I realized.

"Try to be democratic about it, and give both parties equal air time."

"Or all parties—that's the meaning of a democracy," I suggested with a laugh.

"Yeah," Wanda said, "I was just thinking while you were talking about race, how America is supposed to be this wonderful melting pot. And we are supposed to be the place where all men are created equal. And look at us—we still feed on violence and hatred and racism everywhere you look. But America's multi-culture is not about the different races. These people left their countries to unite as one culture to unite under the United States Constitution and the Bill of Rights. These are the documents that give freedom by divine right and that is our common heritage and culture." My sermon had gotten to her, after all.

"And look at us," I added, "Are we really a country of equal rights?"

"Not yet, but I'm hopeful. Our forefathers wrote a powerful document. And they laid their lives on it when they signed the Declaration of Independence." Wanda was starting to wave the flag.

"I've got this other book right here that you must see!" I said as I pored through the pile of books and files on the floor by the fireplace. "Here it is, *The Secret Destiny of America* by Manly P. Hall."

"What's this?" Wanda reached for the book, but I wouldn't let her have it. Instead, I flipped through the pages, looking for something in particular.

"Here it is. America has a special mission. Our country is at war with itself, and how many really trust our own government? This is the place about the unknown speaker who swayed everyone to finally sign." I showed her page 119 and gave her time to read it.

"Wow, listen to this," she said. " 'If the Revolutionary War failed, every man who had signed the parchment then lying on the table would be subject to the penalty of death for high treason. It was late in the afternoon. The chamber doors were locked, a guard was posted and after debating for many hours at the State House at Philadelphia, when suddenly a voice rang out from the balcony whose imperishable speech remains and bears witness to his presence. The delegates rushed forward to sign and when the patriots turned to express their gratitude to the unknown speaker, he was no longer there. He was not found anywhere. How he entered and left the locked and guarded room is not known. To this day no one knows who he was.' "

"Wow! Who was that guy who spoke with divine authority?" Dan said nothing.

"Talk about being committed! You know, one of my relatives signed that day."

"No kidding," I said, impressed.

"Yes. I can't remember his name, but my Grandmother Hopkins studied the family background with a fine-toothed comb. She got me into Job's Daughters because she was one of the D.A.R." Wanda put her nose in the air and both thumbs under her arms as a gesture of pride. I was looking at my neighbor and could not help but wonder where she was coming from.

"What's the D.A.R.?" I asked.

"It means the Daughters of the American Revolution," Wanda announced proudly.

"Whoa! I am impressed. That means she was probably a Republican. What happened to you?" I mocked.

"Well." A quizzical look grew on Wanda's face, finally turning to horror. "Oh my God, you're right."

I couldn't help but laugh. "I knew there was something about you I liked. Maybe there's hope for you, yet."

"Don't count on it. I will never bend on some issues. And being a Democrat is one of them."

"Is there no hope for us, Lord?" My humorous plea was sent to the ceiling, in vain.

"Look at this," Wanda was back in the book, talking about the American seal. "On one side is the American eagle and on the other is an unfinished pyramid. Here it says that's to 'represent human society itself, imperfect and incomplete. Above it floats the symbol of the esoteric orders, the radiant triangle with its all-seeing eye.' "

She showed me the page.

"They probably chose these symbols to represent America as a brand new nation, like the phoenix rising out of the ruins, and guided by God," I speculated.

"Yeah, Benjamin Franklin wanted to use a turkey instead of the eagle."

Wanda's remark reminded me of a story. "I think my brother, Fred, may have first coined the use of turkey to mean a stupid moron when his family was on vacation one summer. He was driving behind a truck stuffed with live turkeys. His youngest son asked where the turkeys were being taken. 'The turkeys are going to be killed, slaughtered, then eaten,' Fred said. His kids thought that it was very gross at first, then Fred yelled, 'But that's everybody's Thanksgiving!' The kids all of a sudden switched gears, got into a festive mood and started saying things like 'Oh boy, yummy' and stuff like that. Then Fred noticed one of the turkeys squeezing through the fenced boards on the back of the truck. It fell to the roadway. 'Look kids, he escaped!' Fred said. The family was all yelling in favor of the turkey to get his freedom from a sure and gruesome destiny. Then Fred yells, 'Look at that stupid turkey! He's chasing the truck—what an idiot.' 'Why's he doing that, Daddy?' the kids ask, and Fred says, 'He's trapped in his own metaphor. This means he's caught between the status quo and his path to freedom. In other words, he is doomed by his own choice and doesn't understand either option. He chooses only what he knows and what's familiar to him.' I had to laugh as I remembered. But when I didn't hear Wanda's laughter joining in, I realized the story was lost on her.

To prove the point, Wanda sounded as if she was just waiting for me to stop talking. "Listen to this!" She was still engrossed in the book on America's destiny. " 'Only enlightened men can sustain enlightened leadership; only the wise can recognize and reward wisdom.' That's sure the truth."

"That's what this is all about. All I'm after is the truth, Wanda. What people want is the truth. What thousands want to know is, is the Earth hollow or isn't it? Does an inner Paradise Earth exist? That's it." I suddenly realize I am tired.

"Why, Dan? Why is that so important to you?"

"Because it's our human birthright to be in Paradise, and if it's a major world government cover-up, I want to know the truth. But, I believe it's even too big for the government to cover-up. It's that simple, Wanda. It's my home."

"Paradise is a metaphor, Dan," insisted Wanda.

"What? I can't believe you have come this far with me and you haven't gotten it!"

Wanda continued, "Well, yes. At least, that's my belief. I guess your understanding of Paradise is different from mine. And Ritter von X's interpretation is another concept entirely. It's so easy to come into a discussion like this with judgment. I can see he is really sincere and a fairly articulate and educated person and backs everything he says with his own experiences. All humans need to have a worldview of their own. Mine just isn't his."

"How can truth be separate from itself?" Now, I was irritated.

I felt like the turkey. We, too, are trapped in our own metaphor because the truth remains beyond our reach. Perhaps to believe in a truth is a trap and knowledge is a higher metaphor to realize truth. "Metaphors can and must change, Wanda. Do you think when a caterpillar goes into its cocoon stage that it knows its true destiny? What I'm trying to say is, does "cocoon-like" mankind knows the vision of his next world—the butterfly?"

"I don't think he does, Dan. He doesn't have the slightest idea of who he is to become. That's just the problem." Wanda is warming up to her convictions.

"Even if you're right," I said, "mankind needs to become aware of his next world. The cocoon is like the Dark Age period until the transformation takes place. The butterfly is the quintessence of its journey. Paradise Earth is the quintessence of mankind. My visionary experiences are of Paradise Earth. Many people remain trapped in their own metaphors, Wanda."

Yeah, just like the turkey running back to the truck to be slaughtered—as if it somehow knew it had only one true destiny to follow.

"Could the Hollow Earth be a metaphor, Dan?"

"Yes, of course, Wanda." I paused. "But I don't think so." I was deliberate. "But if it were, it would still symbolize the vision, and the plan to restore Paradise on Earth. Like all metaphors, they must transform themselves. Even church and state are metaphors, Wanda. Just think, what happens with government when the people find their God? Isn't this what a religious faith is all about—to have a complete fellowship with the Creator?"

"Yeah, yeah, you're right," Wanda conceded.

I am happy with my final argument. Wanda did not begrudge me the point. We both laughed as she continued reading the Ritter von X transcript, more intently now until she reached the end of the page. Then she looked worried.

"Danny-boy, your metaphor about the rainbow is great, but he doesn't seem to get the point, and then you seem to turn around and agree with him."

"What are you talking about now?" I have begun to feel impatient again.

"Well, what do you mean about the Chinese all disappearing, and then you say, 'What about the other races and the black people?' What am I supposed to think?" She was losing control. "How can you sit there and agree with him?"

"I did agree with him—in principle and in truth. But, like you, I was shivering and trembling in my own fear and ignorance of racism. Wanda, please try to understand my vision. I stood as one single eye. I tumbled through the bands of the rainbow. I understood each degree of vibratory consciousness. The Earth and its people are handicapped. All life on its surface is suppressed due to the catastrophic change. Wanda, let's give von X the benefit of the doubt and try to understand what he is speaking about. He's saying that our root origin is in the Hollow Earth. We came from there. Even the Bible says that after the flood, people who lived on the Earth changed. Even language changed into many others, and other races came into being because of the harsh conditions after the flood—after the canopy was destroyed. God gave Noah the very first rainbow on the Earth, as a covenant. Before the flood, there was no rainbow. After the flood, the first rainbow appeared. We gradually began to change, adapting to the new atmosphere and environmental changes. Then newborn races emerged. Yes, we all spoke one language and were all one color. Maybe the original human race was white or black or even blue!"

"Dan, how can you say that?"

"Wanda, haven't you been listening to what I just said? Does it really matter what the hell color the human race was or is or will be? Remember, we are a product of our environment. We've heard about other culture extremists claiming superiority over one another for one reason or another and in the end they always fall. Let me say it another way. All men are created equal except in consciousness. I had a black professor once tell me the same thing himself. He said blacks developed dark pigmentation because of the sun's harsh toll, and it's just a product of our environment. It was an element of pride, for him, because of their ability to survive." I paused and sighed.

"Here's another source that says the same thing." I scrambled through the piles of books in front of the fireplace and find Aghartha, by Dr. Raymond Bernard. I flipped to the page I wanted and began to read out loud:

The blond races are those who more recently left the Subterranean World and lived for the least time under the unfavorable conditions of solar radioactivity, soil depletion, and wrong diet on the earth's surface, while the races that were exposed for the longest period of time to the harmful rays of the sun and general unfavorable effects of life on earth's surface, turned darker, and under the tropical sun, turned black. The deposit of pigment in the skin represents the body's effort to protect internal organs from the detrimental action of solar radiation by forming a protective covering in the skin, in the form of dark pigment.

Similarly, the hair turns black and the eyes brown. All races, including blacks, were originally blond and blue-eyed, as infants tend to be at birth. It was unfavorable environmental conditions that turned the original fair Nordic type dark, as occurs under the unfavorable conditions of the tropical climate.

I stopped reading and looked at Wanda, who was staring blankly back at me. Then I continued, "That's the formula I wrote on the window over there, when I remembered our time in Paradise. That's when I realized the time difference between our surface on Earth and the Paradise Earth. We are seven seconds from our original presence. Our consciousness is caught and held by an altered state of awareness. Change the perspective and our awareness changes.

When awareness changes, our reality changes. When that occurs, things are seen as they actually are. In other words, the way the Creator sees all things and us within in it. This is the grand illusion. As an example, a similar effect takes place when you insert a pencil in a clear glass of water. The pencil then appears bent due to the law of refraction. The same idea is that our consciousness is like the pencil altered by the frequency of the sun and the [water] the atmosphere alters or bends by the law of refraction.

Simply put, all things are not what they appear to be. The consciousness of the human race remains eternally captive, so it appears. This is all due to the cataclysmic changes by divine interference. Hence, we look back at our reflection cast into the vibratory color frequencies and seen as a rainbow. We are the result of our Second Fall, here on the physical stage. That's when we lost Paradise Earth. Man's First Fall. That's when mankind fell from grace—when Adam and Eve—or rather, an ancient civilization that symbolized the first male and female, fell from the cosmic stage."

"So, the world's a stage, after all?" She laughed.

"Right, we have to rebuild the stage set—"

"And go on with the show—"

We both laughed, but I heard a hard edge in Wanda's laugh. I went on. "Yes, if we went into the Hollow Earth, we would eventually change again. Our eyes would become blue and our hair would become blonde. We would become larger in stature and become ageless—again, because of the environment. Our eyes would turn blue due to the intense atmosphere. Our hair and complexion would be a near-white blonde and we would have fair skin. This is because of the soft radiant life-giving inner sun. Our bodies would emulate the light of the interior sun and absorb the radiant energy from the atmosphere and from the life-giving plants. It's simple, Wanda. We are a product of our environment. Not to mention the abundance of high energy-giving fruit and vegetation for food in the inner world," I added.

"If the canopy is rebuilt here on Earth, we would all change here on the surface?" Wanda looked for confirmation.

"Yes. Correct. That's the mission and the plan. The rainbow is a symbol of our bridge to eternity. Paradise becomes our home once again. With the canopy itself restored, we would gravitate to our original forms as we were in Paradise. We would again be in God's Eden and in His image."

"That might take years—or centuries!"

"True, but I believe this is God's true purpose for all humankind. Even the last pages of the Bible proclaim our eternal liberty without death—that we will live well beyond a thousand years and then begin to remember our true origin. We can't look at that as any kind of major separation, but a blessing that a flame exists in the hearts of certain individuals, who have remembered their source—their origin. The enlightened ones always walk before us. They escort humanity through darkness and ignorance, from death to eternity."

"You mean, like intuition? The flame that exists is in the hearts of certain individuals and is like a sixth sense about our origin?" Wanda tries to restate my words in order to understand.

"Von X calls it 'genetic racial memory.'"

"Like Akashic records?"

"This is a dawning of a new new-age technology. We need to understand what our mission is, and we must be able to be wise and judge our direction carefully. Wisdom means to judge all things rightly," I clarified. "Does all this make sense to you, Wanda?"

"Sort of. Are people without that certain 'flame' in their hearts inferior people?" She was trying to resurrect the race issue.

"Unfortunately the word race has negative connotations, somewhat like the word diet. It means something different to everybody. How would you answer if I asked if the cells of the toes are inferior to the brain cells of the same body?"

"Of course not," Wanda said, resolutely. "And, as Jesus declared in his prayer to God, 'Thy Kingdom come, thy will be done on Earth as it is in Heaven,' Earth is the Kingdom in which Paradise will be restored within you and our physical world."

"No, that's not it, Wanda." She was off the point. "Those people, because of generation after generation after generation of forgetting their link, have become a product of another type of creation, another plan. So, they become what I would call the darkness and the antithesis to the plan of Paradise. A great shadow is cast upon us. And we fight many different kinds of wars over that. That's been our human history.

"The Akashic memory or the genetic racial memory of Paradise Earth is either latent or has been obliterated from our internal existence. It is because of those rare individuals who do remember the flame of our origin, and become responsible for administering its meaning and destiny, as I am doing at this moment with you, that there is any real idea of what hope really means. The Earth is our Paradise and not some junk heap or cosmic garbage dump that men are making, and have made it out to be. Just look back at all of the great civilizations that remain as dust in the eternal winds of the earth.

"The Atlantian beings existed in several forms. In other words, there were races that had not been on Earth as known to us today. They had been here before in ethereal forms. The North Pole at one time fostered a highly enlightened civilization, with understanding based on the conceptions of height, depth, and width. The three dimensions existed then and now, but the last fourth dimension is known to them and lost to us. This dimension typified the races at that time. These races each held in them the yellow, blue, white, and green races. But the Golden Race supervised all these races."

"What?" Wanda snapped, clearly confused and skeptical.

"Imagine Jesus as a Golden Age man who taught and performed all he knew, directed by the light of the Creator. Now think of humanity as a Golden Age civilization, people, or race. What do you think such a Golden Age civilization could accomplish besides overcoming death, disease, and war? Would not this be a Golden Age race?"

"Wow." Wanda's face turned reddish-pink and her breathing became short. She caught her head and clutched it between her hands. Then she looked up in a quick motion and said, "How and where are you getting this information? Dan, how can you just go on and on without a written script?"

"It comes from the light, Wanda. Once we are contacted by light, it's enough just to think about forever."

"Are you saying you have the knowledge to rebuild an earthly Paradise?"

"In all honesty, I must say yes—our alien space friends have the technology and want to give it to us." This is the 'secret destiny of America.' The forefathers of our founding nation, knowing this need, have not lived or died in vain; rather, they opened the light of freedom. Freedom is a light and power, a consciousness coupled with wisdom of the ages and love for our Creator and his law. It will cause our souls to ascend into a great knowledge to rebuild Paradise Earth.

"Our world's best minds need to advance into a different type of technological engineering. They need to seek to understand the vast knowledge of the chief engineer of creation, knowing that Paradise can be rebuilt. The human race is becoming aware in a very nebulous way while clubbing themselves with constant debates about environment, alternative energy, the greenhouse effect, global warming, and so on. People react only out of fear, lack of focus, or the moronic idea of carbon footprint world taxation. If you can dream it, human ingenuity can build it. What else am I here for—if not for this? While I sit here in the eternal memory of Paradise I remain an open vessel for its creation. This Paradise for humankind features the most extraordinarily intelligent design that has ever been conceived."

"Wow. Dan, is this really possible?" Her eyes began to swell a bit and tears began to trail down both cheeks.

"See, Wanda, your own soul proclaims it while the angels standing by weeping through us." Wanda fell into sadness, then silence, listening perhaps for her own still small voice within.

"Wanda, the next book I write will be in theory how Paradise can be built, and how to gather the world's minds and create a think-tank like never before in human history."

"Like Peter the Great!"

"Yes, like Pharaoh Ankhaton, who gathered all the most learned individuals of the time and opened great halls of knowledge in science, math, music, as well as vast knowledge of our universe."

Wanda was still wavering. She remained concerned, "Dan, some of this stuff is too sensitive to write about. It scares me to death, as a matter of fact. In principle it sounds great, but if I apply that principle to the race issue, am I a racist?"

"Are you asking me? I don't know. It scared me too at first. But my job is to face all these issues, no matter how scary they might seem."

"And I respect that. What about the soul issue? Does Ritter von X think there are really people walking around the planet with no souls?"

"I suppose those who have no conscience are without souls. Those who destroy humans in hideous ways either have no souls or are agents to a soul less humanoid god-like entity. I'm sure you can identify some who may fit that description," I said. "Only the creator can create a soul and only the creator can destroy one. This is what the Second Death means, Wanda. My personal experience is that they are a judged people—'they' meaning some of those who lived in Paradise and who still live among us today. And, by that understanding alone, it will take a cycle of ten thousand years or more to recapture their original over-soul. They will have no memory during that time—no idea of who they are, or why they are, or where they are going. Based on that understanding, I have to say yes."

"Like the cocoon?"

"Yes. But it was Jesus," I continued. "The mission of Jesus Christ is to call all his flock back. Christ is the 'over-soul' with a peculiar mission that transcends our Earth into its Paradise state. Humankind is the offspring of the over-soul. Only certain individuals will traverse beyond the Christ consciousness and enter the Godhead of creation. Christ is our beginning and ending. But I'm afraid even most who claim to be Christians have a serious and gross misconception of the meaning of Christ. Why do you think many are so-called anti-Christ's? That's why we need to really wake-up because our adversary will devour us.

> *The Bible quotes Jesus [John 14:8-12]: Truly, truly, I say to you, he who believes in me will also do the works that I do; and greater works than these will he do, because I go to the Father.*

His mission is to find His scattered sheep and bring them home again. He calls upon all who can hear to be taken back home. This gives purpose to all mankind. With God, mankind will be used to rebuild the Paradise Earth."

I stopped, hoping she would catch the drift. "This is my experience, anyway. This is what Christianity means to me. God enlightened Buddha and countless others and God can enlighten anyone, but Jesus Christ is the Son of God, unfortunately most have not become conscious of this. I took a breath, and then continued, "Remember when I told you once on the phone that I had the whole race issue resolved? And that God's enlightenment was the only answer?"

Wanda nodded her head—yes.

"We are the sons and daughters of God, and God is to be worshipped as God alone. He answers all our questions in light. Once that light comes in, then we must paint the picture. He gives us the enlightenment—all color in one. We then must take that light, break it down into a color palette, get a brush and a canvas, and paint the picture as the reflection of God himself. However, we are speaking of the picture of creation and the handiwork fashioned by God. We are the result and reflection of His image and intelligence.

But like an unfinished painting, creation is incomplete. And we have to appreciate the pieces, as they exist now, and understand their differences. It is the opposition we must dismantle with compassion; we must replace our fear and ignorance with knowledge and wisdom. If we understand this plan and our differences on the Earth, the doorway to Paradise will open. The plans of Paradise will commence and the rebuilding process will begin by the people of the Earth. I hope all earthly races can accept this as their identity and origin and understand the destiny ahead. Humanity will bridge this great cosmic plan by the power of the cosmic and angelic hosts, our space friends, and bring this great vision into reality. The world as we know it will fade away and won't be found in our memory again. The power will be received as it was in the days when certain individuals were enlightened and held the first blueprints of the Great Pyramids.

I looked at Wanda and suddenly realized she had been following me closely. Her question was a logical one.

"Where does the 'angelic host' live?"

"They live in the center of the sun! Which is the center of all suns as the central sun," I blurted out.

"Dan?"

"It's just a joke, Wanda. But, not really, please remember, appearances are not what they seem to be. That subject will be left for an additional writing at another time."

"Let's don't get into that one!" Wanda laughed. But we were on a roll.

"Is this a metaphor of your belief system or a figment of your imagination?" Wanda asked.

"An immortal has no need for reincarnation. Everything on Earth reincarnates—plants, fish, animals, man, even the geological rock cycle of the earth. Once again, it was due to the catastrophic changes to the Earth. The people of this world are slaves to this system and have been cast out of Paradise due to the cataclysmic event. Many people today worship a false god."

"I think so, too," Wanda agreed. "Some of the world's religions are limiting, godless, and destructive."

I thought for a moment. "Many religions have played a destructive role at some time during history. However, some use brute force and program your thinking. Nothing is more heinous than the Islamic extremists using brute force. We will speak more about that later. Each religion believes it has a vision for humanity. The fact remains that we must receive the power to make that vision a reality or subject ourselves to extreme Islamic world dominance. When any religion claims to be the only true church with radical methods of death to non-believers, it poisons the very fabric of our first amendment rights on which it was founded. Something cannot be a little bit cancerous and some things just cannot be accepted and have the religious freedom and protection at the same time. Such religions are the true enemies to freedom-loving people and I believe one day these types of religions will no longer be tolerated by freedom loving people. What happens to freedom then? Isn't it better to serve freedom than follow the dictates of unenlightened men?"

"Maybe you're right," Wanda said, pensively. "Jesus said, 'Greater things you can do than I, and greater things you must do—' "

"Listen, Wanda," I interrupted, "My mother had a dream once that she spent long, endless days and nights painting the ceiling of the Sistine Chapel that Michelangelo had done. She was painting the last part of the Chapel. The myriad works began to form into a giant, swirling, multi-colored wheel. It was like a large, colored mandala. The colors blended with the images and changed into a giant wheel that began to spin around and around, slowly at first, then gradually, at high speed, until she noticed all the colors were blended into one color, and turned white. She felt suddenly enlightened. 'What's in a color?' she exclaimed. 'It's all white!' She placed her hand into the center of the white, swirling disk of light and felt an endless eternity inviting her into its center. She wanted to add something in the center to complete the greater meaning of humanity. She tried to explain to me the meaning of her experience and what had come from the depths of her soul about the unfinished plan of creation. I told her it was the dawning of the unfinished work that will commence in a later time, when we are more perceptive."

"Exactly," Wanda smiled. "Your mother's dream says it quite clearly and powerfully. Truth is the white light. And that's God."

"Well, I don't have an absolute corner on truth, Wanda. But maybe it's the absolute source that's at the center of all being and creation. And maybe a world exists there." I thought about that for a moment, took a breath, and then continued reverently. "Let's hope we humans have the wisdom to humble ourselves so we can know and understand these differences among all races on this planet. Each avatar has been enlightened for its particular time and need for its people. I believe that Christ's gravest concern is about the survival and destiny of humankind, God's promise of life eternal, and our eternal home of Paradise Earth. And by all means, read the Book of Enoch who describes in detail about Paradise before the great flood. He lived 365 years and then God took him. The differences are not separate truths, but rather, separate identities of the same truth! Maybe that will shed some light on why Lucifer is who he is, and why he fell from grace. And down we went with him! Makes you wonder why mankind turned from God in the first place, right?"

"Probably—because no one has a corner on truth," she said, smiling more widely.

"Exactly, Wanda, and once we recognize the bridge between mortality and immortality and understand the differences, gladly we will leave the old worn-out carcasses of religion, governments, races, dictators, disease, death, metaphors—leave all of it behind in the old world. Humankind is the living tomb of the Christ consciousness. The name we call and refer to as "Christ" will be known to those as the nameless God. But it is yet not time to know. If we can humble ourselves to realize this that makes abundant room for the greater light to enter this unknown domain, this vision of man's destiny can be received in those who, in their hearts, love God, his law, and creation. The human consciousness is a cave within the mind of God. This is the precise place I focus my faith in what's left undone in Gods name."

"Thank you, Brother Dan."

"There's a lot I don't know. But I do know that my faith in my own experiences remains alone with God, knowing all things are possible. At this moment, let us hold in the light of faith what we don't know, while we petition the cosmic and its hosts for greater enlightenment to show us the way. With this greater enlightenment, our Earth will surely see the dawn of a new age. I am surer of this than anything I have ever experienced in this world."

Wanda nodded her head, smiled, stood, and gave me a kiss on the forehead. I have come to a stopping place. I took a bite of German chocolate cake and a sip of my Red Zinger tea. We looked at each other and took deep breaths.

"So, do you feel like we're finished with all this stuff, Dan?"

"We're finished for now."

"Yeah, well, I'm missing Days of Our Lives. Maybe I'll just go on home."

We looked at each other again and took more deep breaths. Then I jumped up with a flash of inspiration.

"That's it! That's it! The rainbow represents our own separation, yet it also represents how we can harmonize it all, neutralize it all, put it all right. The rainbow's colors are not at war with each other. If they were at war, the power of darkness, this antithesis of light, would become our worst adversary. Perhaps Lucifer is a superficial devil. Maybe he's really God, wearing a mask to frighten His creation with fiery power to direct His creation. Maybe Islam is a great antithesis to the rest of mankind and we either find our way or submit, because I fear immensely, we cannot ever co-exist and remain free."

"Dan, it's probably the other way around. It's the wolf who is in sheep's clothing, remember?"

"But Wanda, who do you know in our world who wears garments of sheep's clothing? Humanity itself is responsible for all its actions. But of course, God is ever-watchful at all times. All those colors identify God's unique consciousness, and we have the free will to use them to paint a new vision of the world." I sat down and sighed heavily. "It's so simple, Wanda, it's so simple."

"Thank God we figured it all out."

But Wanda did not sound as though she had figured it out; she sounded a bit baffled. "Eat your cake, Danny boy. Who needs Days of Our Lives when I live next door to you? So, how do we get started painting?" she asked in a sincere voice.

"With love."

"Love?"

"Love is everything—you know that. Love is the window to God. You just need to open it yourself and let the fresh air of light enter into your heart, mind, and soul, and let it direct all your activities. But who is God? You are, Wanda," I winked at her.

"God is love and love is God, right?"

"Love is God's light and power. The Second Death will eliminate all the evil in the world. That's what the Second Death is for. Let God take care of it and implement the Second Death. That's final!"

"When do you think we'll go through the Second Death, Dan?"

"No man, nor Christ and his angels, knows the time or the hour. God alone knows that, Wanda." I said this with conviction and finality. After a moment, I continued, "But I believe we need to prepare and show God that we, as a people who love Him and His creation, will be worthy of His presence when He decides to return. We need to start by dismantling our aggressions and humble ourselves with a nonviolent attitude throughout the world. Maybe even a day or two of fast and prayer throughout the world would create an observance of our Creator—what do you think? We have Earth day, why not focus a day to ponder the Creator?"

"If you can pull that off, Dan, you're one powerful hombre."

"No, not powerful, Wanda, just persistent, stubborn maybe. But, hey, someone's got to hang in there for God, freedom, country, and family."

"And you, Danny boy, are that someone. Catch you later." Wanda winked at me, suggesting she agreed with me. As she made her way from my patio to the street, I could hear her singing in her best George Strait imitation, "When we are all one race, and that's the race of humankind, we shall be free!"

I yelled out the door after her, "Keep the vision, Wanda. Help fulfill God's promise of Paradise. What are words without action, but just words? This is a new Exodus, Wanda!"

Chapter Ten

❧ Ancient Memories ☙

I want to know God's thoughts; the rest are details.
—Albert Einstein

There is a huge oak beside the Shell station at the Highway 17 turnoff onto 680, north of San Jose, California. It comes alive in the dusky light of early evening. Cries of a thousand feeding sparrows searching out the final worms of the day can be heard. A low rhythmic inner sound became audible as I was filled with a vague memory that came into slow focus. In the memory, my brother, David, is pulling up to the curb to take us to his home for tomorrow's Easter dinner with the family. Dave's wife, Sue, was also in the car.

"Come on, Danny, get in," David urged. He was in a hurry.

Katrina got in first. Silent in the back seat, we rode to David's home. A strong image came to me, focusing into a message, which then became audible to me. Why don't they hear it?

On the surface of my conscious mind, then, I could almost feel it. Everyone should know. All of humanity must know what I was hearing at that moment. Did David remember why we were here? Surely, he had not forgotten what we came here to do. The Great Plan, yet unfinished, had been handed to us since the beginning of time. Surely, he must remember.

Surely, he must remember our conversation when I was just thirteen. "Remember the dream, Dave?" Our words are still clear, sharp to me—transformational for us both.

"Oh yeah, you mean the one about the great white circular light?" Dave's concern for his little brother was genuine.

"Yeah, Dave, we're supposed to do something. I know we are. I mean, it's actually taking me somewhere. I get swooped up, and then

I can't remember anything when I wake up. But, I don't think it's a dream."

"Danny, listen, maybe it's not a dream. Now, I don't want you to become afraid of this thing. That's the worst thing you could do. It may be trying to tell you something. I don't know what it is, but I do think that we—you, Fred, and I—are not just like everyone else. I mean, I think we may have some purpose here that we don't really understand yet. I believe it's some unknown mystical destiny." Dave was twenty. He knew what he was talking about.

"What do you mean by destiny, Dave?"

"Oh, geez. Okay. Destiny, something special to do, you know?"

I felt better about the white light just then, without understanding why, and that conversation nine years previous had been preserved like a moving picture in my mind from that moment on.

Driving into San Jose, my thoughts reflected off the windshield like a mirror. How could they not see what was going on? A new tomorrow and a promise from God, vibrantly apparent right there in the windshield and in the lives of every one of us.

While I am in harmony with the cosmic, of course, I was also in the car with Dave, Sue, and Katrina. Grounded in this world, I spoke.

"How many are there of us tonight?" I asked Sue, not knowing why such a strange question came out, or why it was directed to her. Yet I knew that her subconscious memory was in tune with this unknown Great Plan.

"There are tens of thousands of us," Sue's answered, unhesitating. She was aware.

I was reassured and I rested easily with a song in my mind for the new day ahead. I began thinking of how the angelic beings dwelled within the hearts of men. Their destiny was to lead humankind into the symbolic meaning of the seventh and final day of God's plan and complete our ascension. Once this was accomplished successfully, God's earthly Paradise and Kingdom would be established. Humanity's mission upon Earth would be complete. With those thoughts, I fell asleep.

The car stopped in front of Dave and Sue's house. I woke up and we all went in. David prepared some raspberry leaf tea while Sue and Katrina chatted.

At that moment, it seemed strange to me that gas had to be employed as a fuel, since all things that came from the Source of the Eternal were for our material use. Why did they have to be changed, broken down, consumed, transformed, and then taken again into our bodies to be used? It was a repetitious and continuous cycle of meaningless life and death—all for what? How barbaric. We are all just a piece of unfinished work. Water for tea, not yet come to boil? When were we going to be done?

"Tea, Danny?" Dave handed me the steaming cup.

I followed him silently into the dining room, where his typewriter, books, and the papers of his master's thesis on the Nuremberg trials lay scattered, unfinished, and unshaped on the table. I walked through the veil of steam from my teacup. *When are we going to be done?*

"Danny, are you okay?"

"Yeah, yeah." I was aware of his concern. I forced my attention away from the vague and, with matrix memories of God's great plan veiled in Dave's master's thesis of Nazi Germany, directed my next words into his left eye. "I am just fine, David, just fine. Now, I am going to go upstairs and help my wife make the bed. Thank you for the tea."

He probably thought I was on drugs of some kind. I was not on anything. Something was "on" me. It seemed so obvious. That was the problem. Why did he not understand? He said nothing and only stared at me.

I simply repeated, "Now, I am going upstairs to help my wife make the bed."

His stare of amazement drilled into my back as I left the room.

After we ate, Katrina and I prepared for bed. Standing before a mirror, I commented as I gazed into the reflection of her blue-green eyes, "Our eyes are the eyes of one." I sat on the edge of the bed and watched Katrina undress. The room was dark, but remained half-lit by the streetlamps outside. We lay upon the covers of the bed. I picked up her hand and symbolically removed her rings from one hand. Then I proceeded to the next hand and repeated the same motion. There were no rings visible, only the ones I remembered her wearing in an ancient time, before the destruction of a mighty empire.

Ritualistically, I lifted her left foot and again removed all the rings and ornamental jewelry from her toes and ankles as a way of breaking the karmic memory from the old world. A mystical voice rang in my head, uttering some lost and secret prayer. Deliver her from her foes, I thought.

A voice rang out, *"Here begins the praises and the glorifying of coming out from and going into the glorious Neter-khert in the beautiful Amenta. Come forth by day in all the transformations that please Him! Sit in the Sikh hall! Come forth as a living soul!"*

Deliver her from destruction! Guided by the voice, I scanned her nude body upon the bed. She wore only a hairpiece, secretly and cleverly tucked into her hair like an ancient vestment of disguise, in an attempt to conceal some shameful identity. I put my hands behind her head and asked her to remove it.

Unhesitatingly, she did so, her eyes asking the unspoken question, *how did you know?* She was fascinated and yet perplexed. We embraced. I felt the sun's shadow fall upon the bosom of the moon, and heard a quiet whisper, "Yea, though I walk through the valley of the shadow of death, I fear no evil…"

A force directly from the planet Venus surged through me, strange and beautiful; it encircled our oneness like a huge but harmless serpent. Enraptured, I was also captured by it. We flowed with the force field until the planet's influence suddenly caught hold of me, and as we made love a changed consciousness was created. A cosmic peace of mind settled within us while we felt the rotation of the planet in a slow rhythmic movement through our united bodies. Our single body was the planet's body and its life force. I could see the interior, and the greenish atmosphere was vibrant and filled with energy—energy we had just ascended through by the power of our oneness and lovemaking. With complete satisfaction, a warm infusion of pure knowledge and power filled my mind and body. This was followed by a deep sleep, which fell upon us as peaceful, pulsating, and warm radiation that soothed us throughout the night.

It was Easter morning. I awoke early, before Katrina, and went downstairs. David was busy at the typewriter, nearing completion of his thesis. He turned and gave me a quick Easter greeting. I left the room so as not to disturb him, and walked into the front room. Lying on the coffee table in front of me was a book called *Atlantis*. Picking it up, I traced on its front cover the outline of the continent pictured there. I began to remember things. My mind became flooded with a time when I was free within the light of the Creator.

Suddenly, I dropped the book and ran outside. Katrina and I had shared an Atlantian lifetime! Time was all one there, and we had lived in a perfect state of Paradise on Earth. It once existed; I was certain of that. A garden state of awareness, maybe, or perhaps even a real Garden of Eden once existed, where we originally lived and fell from our cosmic grace. I was sure of it, and I believed we could still live there in our hearts forever, if we should choose it again. We had only to find the key. I flashed forward to our present life when we had first met. She had been wearing an Egyptian Ankh necklace. Immediately I was jolted by an ancient memory of her. That key became a doorway, unlocking my memory from my mundane thoughts as I entered into the great domain of my cosmic self. I was entranced by her eyes and instantly fell in love with her. The memory was like a gigantic wave that came from the past and pushed me into her sight as I began to awaken within her soul.

In awe, I was determined to find that key—that state of mind and spirit—and use it to unlock the cosmic door of awareness into that lost Paradise, the Golden Age. Would it be Katrina who would show me the way, or were we merely destined to live, die, relive, and die in cycles of the same old destruction again and again? That was our choice to be made. *No! Never, again,* I thought.

I was driven to celebrate my rebirth and resurrection. I ran into the backyard, discovering within myself a new way of looking at our world. A new vision had emerged with the dawn, and it was Easter morning! I was filled with a supreme, vast knowledge. Everything I looked at was radiating with the glory of white brilliance. I was bursting with the abundance of life. I saw two large cherry trees covered with blossoms, honeybees swarming. The bees made an unusual harmonious vibration that echoed throughout all creation. It was the sound of life—of the Creator at work. It was the eternal movement of the cycles and seasons. It reminded me of Katrina. She was like a song that lit each day and reminded me from whence I came—our Creator's house. A sound of blended harmonies came alive in me. It was a pure sound that danced like a harmonious song in all living things. It was the secret vibratory name of the Creator, who had made our flesh in his image within a Paradise that blessed man and woman at the beginning of our existence.

I wanted to share this glorious Paradise with my new bride, who had come to me from the very place where this inner celestial music was being made. I ran back into the house to tell her about these profound realizations. Shadows of different times emerged and fell before me like heavy drapery. I entered the temple of King David, the founder of the House of Israel. My memory was caught between a past and future of a New Jerusalem, the great City of Peace. Just as suddenly, time itself was literally falling. Everything I looked at was in shadowy disarray.

Not even the mighty temples of the world could keep the Creator upon the throne before all the dark forces on the Earth. I was tumbling through these shadows of time. Frantically, I began searching for Katrina. I opened the doors in panic and felt a large sphere of emotion invade my being, like the sun's rays when they first touch a newly created planet. I asked David where Katrina was. He just shook his head; he did not know.

A quick fear tugged. Had she disappeared into the shadows? Might she have been swept away by the karmic winds of the past? Why? Did we know too much now? Was I alone? Would I never see her and know her again or share with her the glorious Paradise of the Creator? I stood in the light of the presence of the Most High and realized that I could be consumed by it. Please—not yet, I begged, please not yet. No—not again.

A sudden jolt, like an almost imperceptible earthquake, and I was suddenly in a dark time tunnel, being dragged into April of 1984. I moved from shadow-to-shadow in a long, snaking tunnel, until finally I found myself sitting at my dining room table, listening to a small cassette recorder. As I listened to the guttural and raspy German voice,

I still moved through the shadows of time itself. The voice faded in and out of my awareness.

We have to say, what is the Grail? It is not a golden chalice. A cup or whatever, as fairy stories would have you believe. The Grail is a machine. It is a machine that contains most of all the knowledge and the history of the universe from the beginning. The Grail is here on the Earth, hidden away. When it is finally located by a certain few, and it is activated, then will come all the history—everything that ever was—from the beginning. This is the Grail.

Just as quickly, I was shaken back into my brother's house, then forward again to 1984.

Was that David? No. My head was in a swirl of fear, and momentarily blinded by the morning sun coming through the living room window, I listened as the now-familiar voice continued:

The Grail will change the very spirit of mankind, where he will never ever be the same again. He will be a completely transformed being. The Grail exists. It is the source of all knowledge. It is similar to a generator of some type, like people sitting around in a circle touching hands while receiving a magnetic field of energy. It is an energy that is emitted from the Grail itself that will open the eyes of those who have really, truly sought, and seek to see. The knowledge will come forth to them and will be retained by them. It will be passed on to all [who are] deserving of this knowledge [and it] will enlighten man of his once glorious past. He now will become in total harmony at the central core of the universe.

Yes, there are beings out there. We are not unique. We are not alone. And for man to even begin to think he is alone is ludicrous, bordering on insanity. No, he is not alone, but he is of such an order that he will not fit into the family of the worlds until he receives transformation from the Grail itself. All of this is based upon most sacred signs that astound and otherwise completely amaze. There is no end whatsoever to what one can feel once one is exposed to this Holy Grail.

The Grail is a crystalline generator that contains the knowledge of the universe itself. The Holy Grail is a source of spiritual life and knowledge.

This knowledge, the Grail—heir Danny, contains those things that the hearts of modern man cannot conceive, nor relate to in time as we know it. Every soul receiving this information from the emanations of the very Grail itself, hearts will be ravaged with joy and bliss, and in that moment they will simply be transplanted from the earthly plane to the celestial. Because, you see, the supreme experience of the Grail is actually the vision and union with God. You will discover the secret that is hidden from mankind, and [you] will begin to know the vision of God himself. And, if you will again come to understand what no eye has seen, nor ear has heard, nor the heart of man conceived, you will see then what God has prepared for those who love Him.

Dave was still typing the last trials of Nazi Germany, "Have you looked in the bedroom for her?"

I left him without an answer and went directly to the bedroom. The shadows of time fell before me. She was being kept and concealed against her will, lost in the illusion of our present world.

I found her crying, huddled in the upstairs bathroom, hiding from some evil beyond her control. She was encircled in a blinding red aura. It was pierced by white-hot shafts of light, streaming through the skylight in a futile effort to penetrate her darkest fears. The memory of Nazi Germany had a stranglehold upon her beauty, purity, and innocence. Taking her by the hand, I led her from the room and away from her memory of a tortured past. As we walked from time through time, we finally arrived in Dave and Sue's front room. Once again, Katrina's face regained a gleam with the same aura of the white brilliance I had viewed outside.

My memory was keen. My senses were sharp, even through the dark matrix—consisting of multi-dimensional shadows of karmic memories that eclipsed our being. We sat comfortably together and rejoiced in our ancient reunion with a flood of tears.

Gently, with my first finger, I made the sign of the cross upon her forehead to signify and seal the final destiny of God's great plan. We spoke no words. The cross I had made upon her forehead was a covenant I was making in the presence of the Creator. This was a new beginning to execute the great cosmic plan. It represented the final seal and signature of our cosmic conscious participation and our responsibility to it. I was sure Katrina understood me and the great cosmic plan, and our place in it. It was the first divine step to Paradise. We were about to enter into a world where both peace and strife dwelled harmoniously.

David announced, "Why don't you two just stay here today, Danny? Enjoy your new bride, get some rest, and we'll call you later on after dinner, okay?"

When I gave him a questioning look, he spoke his real feelings. "To tell you the truth, Danny, you're sounding extremely incoherent right now, and I really don't think you should be around people today."

Deeply disappointed, I nevertheless smiled at him as he walked outside and shut the door, leaving without my new wife and me for the Easter dinner Mom had planned for family, relatives, and other friends. I was not truly understood.

Katrina drew a bath for both of us, creating a mountain of suds, and we immersed ourselves in the hot water. We spent what seemed to be an eternity in the joyous moments of our bathing. It was a symbolic cleansing of our karmic history. The water was a form of liquid light. It was an attempt to wash away our treacherous memories of the past and prepare to conquer our last enemy, death. As we stepped out of the tub, our first true steps were taken to that promised journey back into our Creator's home.

I would not have known the oneness with the world if it were not for the presence of Katrina. Yet it was our hope to overcome our world karma and rejoin the over-soul within the heart of the Creator. Once again, the memory of the house of King David fell before my mind like a cloak in time and locked me in its past.

Katrina and I took a walk outside. The light was like a golden hue of peace that pervaded all life and things. The sun was a rainbow entering our hearts and magnifying our love. It radiated and hummed within us as an eternal benevolent life source and power. The ancient Egyptian symbol of the feather—pronounced *croma*—entered my mind as a living philosophy from an ancient teaching.

We walked half a block then returned to the house the way we had come. Katrina prepared dinner, and I made myself comfortable. After dinner we watched television without really focusing. I began to explain to Katrina some important parts of cosmic consciousness that were unfolding and that would continue to unfold. She nodded her head and said, "Yes, I know, Danny, I know. It's okay." I thought she understood me.

Intuitively, I knew she was thinking about her friend, Peter. He was her former boyfriend. She was surprised when I told her I knew that her thoughts were about him. Her thoughts and attention seemed to slip from me. I told her not to be troubled—nothing would happen to him.

I reassured her again, telling her to call him if she felt the need. She was somewhat comforted by what I said, but denied needing to call him. Yet, I knew she really did not believe me or have a real understanding of what I had just been telling her. She became very uncomfortable with me knowing her every thought.

Within the hour, my brother Fred called. He was at a local radio station, where he said that an interesting surprise awaited us. He and several other people were celebrating Easter and broadcasting their message of celebration. Katrina and I waited for the broadcast. We turned up the radio and heard those singing humorous songs and congratulating us on our wedding vows. Their cheers and celebration filled us with joyous laughter. Later I began to feel the meaning of the resurrection of Jesus Christ. Fred was always on the outer fringe of my consciousness.

The next morning, David and I went to the Shell station to determine the cost of repair to my Volkswagen. The service station attendant told us that the rear right wheel had welded itself together somehow, and that he had never seen anything like it before. He fixed what he could of the brakes, and David prepared to tow the car to another shop. When the attendant returned my Shell credit card, I signed it, and along with my signature I wrote a large number seven and crossed it. This signified the symbol of the long road ahead of us. I circled the seven, handed it back, and reminded the attendant of the seventh day of creation. I was communicating in symbols by a subtle language of the Creator. I was speaking directly to his subconscious and at the same time communicating in a common language.

After we towed my car to another garage to be fixed, David drove Katrina and me to her mother's. On our way to Watsonville, images of time raced before and after me. I was silent, seeing the road before us like a path of a large cross. Once again, an ancient language related by symbols spoke to me in silence. Looking down the long highway ahead, I saw a glimpse of my immediate future come into focus.

Another woman would someday influence my life and advance my ascension and the meaning of the New Jerusalem. She would be my future wife and helpmate, and together we would do our part in the coming of this New Age Kingdom. However, an important cosmic path had been predestined before Katrina and me. Our relationship would be a karmic responsibility tried by severe tests and trials that would befall us both. We were destined to fall. This would allow us to understand the great humility and sacrifice needed for the ultimate and divine purpose of initiation into a greater light.

David let us off at Katrina's mother's house and returned to San Jose. As I approached the white picket fence surrounding the house, I saw a white stone. The stone was locked within its form, an energy radiated that pervaded all matter. It was everything that existed within our physical realm. I had symbolized my brother, Fred, as this all-pervading energy. His thoughts seemed to be one with the physical universe and yet, restricted by it. Then I saw the sun. The sun to me was a symbol of the eternal, yet its shape was a superficial one in our world of senses. David reminded me of this cosmic solar force that was restricted by its superficial shape in our minds. I began to see the wisdom and folly in all things. Within him, I saw the inner workings of our solar system. The appearance itself was a limitation. In other words, we see but we know not what we see.

When we entered the house, I met Katrina's two children from a previous marriage for the first time. The first question that came to mind while waiting to meet Katrina's parents was, Who am I—just a man?

Then I met her father and her mother. We began talking. Her mother soon took over and dominated the conversation. I could see her inner godlike self. She spoke to me about world history. She was challenging my knowledge, but in a friendly manner.

In the course of a couple of hours, Katrina's mother introduced me to all her favorite world subjects. She was immensely knowledgeable; she channeled the age-old struggle of man. She reflected with clarity the universal concept of man struggling with his God, seeing the deceiver as himself. It was through her that I could feel and remember humanity as a God race of beings. Then she spoke of Adolph Hitler, telling me about familiar speeches she remembered during his rule, and said. *"We are right, always and absolutely right and right in the eyes of God!"*

Later, Katrina's brother, George, offered us a ride to my parents' house. While driving home, George noticed he was nearly out of gasoline. For some reason, I felt angry about his ineptitude for not having an adequate gas supply. I did not take this lightly. It reminded me of the impending global energy crises that could escalate into a major economic world war. My mind became flooded with the idea that energy would be a universal problem in both our nation and the world. We began to search for a gas station and I grew increasingly short of patience. In a cosmic sense, I realized that many people were unconcerned about this particular issue. In my mind, I began to analyze the problem. But the only remedies for this urgent global problem seemed to imply several kinds of world catastrophes, the major ones being world war, dictatorial governments, and the loss of human freedom.

Conscious of these worldly events, I became short-tempered; I took my anger out on George—wrongly—and began dictating each direction of his driving. I was silent about the cosmic message but angry over how ignorant people reacted to the inevitable dark events that would soon befall and enslave us all. It was my conscious duty to sever and radically strip away the ignorance that holds men in bondage. For this reason I was angry, and I displayed my emotion in an admittedly outrageous manner. Katrina reacted angrily toward me but restrained herself from an all-out emotional display.

We pulled into a Shell station in Santa Cruz, and Katrina went to use the ladies' room.

She had actually gone to call Peter. From a long distance I overheard her saying, "He's really acting strange, Peter. Get over here and see if you can talk some sense into him. Please! He's scaring me!"

I felt man's ill-fated direction ahead, and despite his technological advances, I saw that he was leading himself to certain destruction. In my mind's eye, I saw there was no time to lose for what man must yet accomplish for his survival.

Coincidentally, a few minutes later, my father pulled alongside the gas station. How had he known we were here? An indescribable anger began to resurface in me. An age-old grievance from an ancient Egyptian memory came into my awareness. Suddenly, I saw Peter standing a short distance in front of me. My father, like an ancient warrior king, vowed to protect his eternal sons.

Peter wore black gloves. His hands were clenched into fists. Automatically, my being rose high into the air. A force focused in my mind's eye and that force directed itself toward Peter's forehead. I saw a swirl of intense golden energy approach him, and in a barely audible voice I said, "Back, back! Approach me not!" He stopped his sudden advance and stood frozen in all time. After that, I did not hear of him until a couple of years later, when I learned he had developed a brain tumor near his forehead. His family believed he would die. I have not seen him since.

Meanwhile, my father looked out for my interest and was extremely angry at Peter for interfering. "Don't worry, Dad," I said. "He will never be able to come between Katrina and me for any reason. He has been sent back to relive an ancient memory that may take several lifetimes to overcome."

Everyone just stood in silence and stared at me, confused and frightened. Slowly, we got into our separate automobiles. The memory of Egypt faded. No one spoke during the twenty-minute trip as we headed up Ocean Street, turned onto Graham Hill Road, and made our way to my parents' home in Felton.

Chapter Eleven

○∾ From Out of the Depths ∾○

Out of the depths have I cried unto thee, O Lord.
—Psalms 130, v.1

"It is my intention, ladies and gentlemen, to explore with you the legend of the Hollow Earth with all the accumulated information, modern and contemporary, from Arctic explorers who have traversed the regions of the North Pole." My mirror image reflects back my words as I rehearse it one more time.

"Not quite. I should say, 'Arctic explorers who have traversed the regions beyond the North,' for, as you may be aware, the theory suggests that, in fact, there is no North Pole at all. Right?" Satisfied with the words now, I worked on the delivery, emphasizing various words differently, pausing, and picking up my pencil to point to my imagined audience, dropping my head between phrases with a slight smile.

"Then, once everything has been gathered together and all necessary preparations have been made," I continued, "we will meet in Washington, DC. There, we will conduct an in-depth briefing, and set off for 80 degrees north latitude. This is where, hopefully, my extensive research will enable us to move safely beyond the posted area. To be frank, we may encounter danger. The charts clearly state that 'Craft entering this area may be fired upon without warning.' Why? No one knows. But those of you who choose to join us must know of this mysterious and ominous warning. You will serve as crew members, explorers, adventurers, and yes, maybe even tourists into the unknown realms of the Nordic beings who dwell in the center of the Earth. That is, if, indeed, we make it to the center of the Earth at all."

For a long moment, I reflected, staring into my own left eye in the mirror. "Should I tell them now?" I asked myself. "Or should I wait until we get to Washington?" Telling everything is always risky. I remembered. I lost Katrina that way. As deep as her understanding was, as much as we had studied together, she was not ready for everything. She had been driven away. Maybe I will not say anything about it just yet. But I cannot deceive anyone, either. My mind drifted back to April of 1972, when Katrina and I were nearing the completion of a Kabala class at the Rosicrucian University in San Jose.

While I slept one night, a tremendous explosion erupted in my head, like a great and mysterious conch shell opening from this escaped an array of brilliant, golden light. A sense of beauty and love flooded my consciousness.

The awareness of my physical body was lost in the eternal brightness of the golden light. While the light shone blindingly around my face and head, I said, "This is the light of Jesus Christ." The moment I finished speaking the light disappeared. My words confirmed what I had experienced as another truth.

I was immediately thrust forward and completely immersed into a dark, vaporous atmosphere. I had no human form. A sphere appeared to me of such beauty that the glory of this experience remains indelible. I entered this sphere of beauty, calling out to my brother, Fred, and to my Kabalistic instructor and Grand Master of the Rosicrucian Order from Germany, Erwin Watermeyer, but I heard no reply. When I returned to my familiar human form, and my normal waking consciousness, I tried to explain this indescribable beauty to Katrina and my family. It had been like dwelling in the center of the sun and discovering it too was another great hollow sphere inviting me to go beyond all human understanding.

Approximately five months later, I awakened during an afternoon nap. I was raised up from my body, remaining in a horizontal position. I heard a very loud sound.

It seemed as though I were inside a giant high-voltage transformer, hearing a magnification of the sort of hum one hears from a telephone pole on a still night. The sound intensified and a being appeared, radiant and surrounded with a great brilliance of white light. The intense glowing being stood at the foot of my bed and glared at me. I was too overcome to speak. His radiance was so bright, I could not look directly at it—his illumination was blinding. Then, fading and shrinking, he disappeared.

These experiences taught me how to overcome many fears of the great unknown. Gradually I learned how to have confidence in my inner self. Because of these direct contacts, my outer self began to gain the respect and confidence of the master within me. During the following five months, Katrina grew in awareness and confidence right alongside me.

One day in September, she came to me.

"Danny, listen to this!" She was in a state of astonishment. "I was sleeping, and all of a sudden I was raised up from my body, just like you've been telling me! Then, listen to this," she was radiant as she spoke. "I saw three gold pyramids slowly rotating. From the peak of the middle pyramid, a white light shone on my face and a voice spoke. Danny—it was a real voice!"

"What did it say?" I asked.

"It said, 'I am Ra,' and then I saw you walk along the light toward the peak of the pyramid. I called to you but, without hesitation, you continued the climb." Katrina's utter amazement at this experience inspired her to paint the scene using watercolors. This experience alone led her to research the mysteries of the Egyptian pyramids. She also discovered a latent talent for painting.

In Katrina's dreamlike state, her recollections of past-life memories were, unfortunately, largely destructive in nature. She remembered being destroyed by volcanoes and earthquakes in Pompeii. Her waking consciousness was always in a state of fear of tidal waves, earthquakes, volcanoes, and other cataclysmic forces of nature. She never liked to fly and was riddled with other shadowy fears.

These fears clearly suggested to me that deep in her being she had a vivid memory of great destruction. These memories would surface in dreams and occasionally in her conscious waking state. Our ability to communicate concerning these experiences began to falter. We became so much a part of our ancient past that we found our marriage in serious trouble; I did not know what to hold onto except divine love as I knew it. I knew we needed to be grounded, but I was not sure how to do it.

We were so deeply immersed in our subconscious memories that at times it was impossible for us to find our way through the negative and destructive forces our souls had fallen prey to.

Two days later, while I slept, my brother Fred made an astral visit. He was floating above me and told me to come out. I began to move out of my body. "Hold my hand," I said, "and I will come out with you." My body hummed as I passed through a field of intense vibrations into a nebulous atmosphere that reminded me of myriad asteroids passing behind me. I moved through a negative field of energy into a more positive and peaceful vibration. Fred and I floated in this tranquil atmosphere for what seemed to be hours. He began to spin me around. My consciousness began to whirl in an unimaginable high-speed elliptical rotation, growing into a massive rotating spiral nebula somewhere in the mass regions of outer space. Whirling at an impossible speed, I was cast into my body like an open tomb. I felt an expansion of the universe and then a collapse into the micro world of self. I became conscious in a way I had never known before. I became as a live breathing universe lying in the form of man upon my bed. I was awake and conscious, yet I was not my usual mortal self, but a superhuman being. The "I" that I felt was an immensely powerful, destructive deity. I lay awake in my bed, unable to move.

I found myself speaking in a voice, but it was not my own. "I destroy," the voice said.

"What and whom?" Fred asked in a fearful voice. I possessed a causal authority to destroy the Earth's physical plane. I was a causal deity of some kind with a mission of destruction.

Immediately the wraithlike figure of Fred appeared again, racing through the walls of space and time and into my bedroom. He quickly altered the course of this omnipotent, destructive deity that had taken on massive proportions of power. I was literally sent back to a deep sleep within the subconscious cosmic mind. This omnipotent consciousness was formed deep in outer reaches of space, and then dropped into my mind and body of human will and desire. My normal waking consciousness returned with a great sense of relief and exhaustion. I phoned Fred and discussed the many facets of consciousness within the higher self. Because he had extensive knowledge and experience in these matters, he helped me to understand how I had been awakened within my higher self, and how I could conquer all illusion. He knew the occult secrets of raising one's consciousness.

"I was a man with no head," I said, "and with no boundaries."

"Good," Fred assured me. "Use that. There are no boundaries as long as you continue to claim only the ineffable and eternal existence of God in man. You are His instrument. I'm like a simple brick mason using the tools of my trade to build and shape the consciousness from God and to aid His great will. At least this time a cataclysm upon the Earth didn't occur. You were like a tornado that nearly touched the surface of the Earth. This is the Masonic meaning of the "square and compass," when placed together, unites you with the consciousness of God that regulates all actions of life. The cosmic sub-consciousness is the chief engineer of the universe and you are God in the making. This is the meaning of enlightenment in Masonic terms."

After our conversation, my experiences began to intensify and present greater challenges. The more I battled, the more battles were put before me.

A few days later, in a past-life state of awareness, I found myself speaking with a well-known god representing the soul of humanity in ancient Egypt. In that dream-state, I recognized Ankh-Aten, as a part of my memory. A strikingly beautiful dark-haired woman led me into the past, showing me different temples and a written script. She showed me the stone-carved walls in hieroglyphics, and then read from them. I realized in this dream-state that Aten was the name I had worshipped and adored, once, long ago. Suddenly, I was awakened from this akashic dream-state by a long, soft, feminine musical tone that lasted for several seconds.

Later the same week, after reading late into the evening, I fell asleep, whereupon a magnetic force began pulling in front of my face. It was rhythmically exerting power at my chest, literally pulling me involuntarily out of my body. When I was completely out, I began to view my room as if I had passed by it for the last time. I had a sense—a sensation—of never returning to my familiar surroundings.

"There are no boundaries," Fred's voice echoed in my head, "only the ineffable and eternal existence of God in man."

I awoke to the awareness of my true source, slowly and carefully moving around the room in a ghostlike fashion. I centered myself in a firm position and breathed deeply, exhaling directly into the magnetic force and speaking boldly to our world of illusion. "There is but one God!" I intoned, and was satisfied with what I had said.

Once more I took a floating circular stroll around the room. I had arranged it—systemized it in perfect order. All the objects in the room were subconsciously alive with familiarity and warmth. I hovered in harmony with the room's contents, dispelling the pull from the unseen magnetic force. I quietly returned to my body, relaxed and content with what I had said and done. My memory overlapped my objective consciousness and exposed a twofold pattern of time. The past was simultaneously living and blending with the present.

Several days later, an intense dream ignited a series of strange, mystical experiences from underneath the ground. They led me directly to my first journey to Iceland nearly ten years later, and the subsequent search for the center of the Earth.

In that first dream, Katrina and I had rented a house that was haunted. I went downstairs and entered a rustic room, the walls of which were old gray wood nailed together in irregular forms. Bright light shone in all directions through gaps in the uneven paneling. Katrina did not want to go there with me. The dirt floor had old boards stretched along it for a path. Extending upward were crude wooden stairs that creaked as I entered our bedroom.

The next dream of that same series was more significant. I was driving down a rocky road between wide-open lands in Iceland. It was early evening. A woman passenger was showing me the open countryside of Iceland. We approached a distant farmhouse on my left.

I stopped at the entrance of the long driveway and stared at the farmhouse. Without hesitation, I drove up the driveway, recalling its familiar surroundings. Excited, I walked quickly to the door and knocked. A tall, middle-aged man answered. I asked him to take me to where the crystal was hidden. He immediately agreed and invited both of us into his house. He asked us to follow him downstairs, where the crystal-grail had been located for several thousand years. He said that no one had been able to understand how this device worked.

The crystal-grail stood about six feet tall and two feet wide. I was enthusiastic and consumed by memory. I told him that I had designed and built this particular crystalline generator device and I remembered everything about its construction. Its purpose was to protect the Golden Age from outside forces of anti-matter. I ignited the crystal by tracing a geometrical pattern with the index finger of my right hand. Tracing that particular pattern over the crystalline facets caused the crystal-grail to implode with a gleaming white light. The light encompassed our bodies, and then penetrated deep into our minds and souls—until everything was formless, but being one with the white light source. It was like a gentle blinding fog that obscured our vision until we could no longer see.

Instantly, I found myself moving back in time. I was thrust into a tremendous battle; heavy metal swords clanged while heads were being severed in a bloody battle for the crystalline-chalice. Victory was ours when all heads had been severed from the men we fought. At the moment of our victory, I awoke and returned to the present.

Another dream also involved subterranean images. Katrina's son, Art, and I were somewhere around the interior of Iceland. We were exploring and diving into a large round crystal-clear fresh-water pond. The water was warm, heated by underground volcanic activity. We dove into the water and saw a brilliant bluish-white light in the depths of the old volcanic pond. The closer we got to the light, the brighter the water became, until the water itself became as liquid light. We swam closer and closer to the source of the light, touching its source, then surfaced and entered a large cave that extended deep and long underneath the volcanic underground cavern.

Inside the cave were many large stone-carved statues. Each statue was like a mighty god or conqueror. It was as though we had entered the secret dwelling place of the gods. Each figure portrayed its might and power. It was like viewing the history of man's consciousness from the beginning of time to the present.

Suddenly, a strong, bold voice spoke, "Why have you entered here?"

My step-son and I looked at each other while standing in the midst of these ancient conquerors, and I replied, "We have come here in peace. We mean no harm."

Then, as the bold voice spoke through one of the mighty godlike statues, it came alive and approached us. "If this be so, then you must pass through me."

Art and I looked at each other again.

We began to walk toward the speaking statue, but it seemed to come alive and blocked our path by expanding itself as large as a mythical giant. It raised itself high beyond the earth, its head seeming to turn into an unknown inner sun and its body appearing as enormous as the distance between the Earth and the sun.

Art and I looked at each other again.

"We can equal this and more," we communicated without words. And we became as one, rising beyond the size of the deity that had challenged our sincere offering of peace.

After this dream, I regularly began to hear eerie sounds, ghostly "ooohs,"—almost cartoonish in nature; they were trying desperately to terrify me. One early morning, as I was eating a piece of buttered French bread and gazing out my window, an apparition suddenly rushed at me, grabbing as if making a single fierce attempt to seize my bread. I wrestled with the ghost, sensing a life and death encounter, even though it seemed only to be after my bread. Grappling wildly with it and grasping for my bread, I managed to get it outside the room where there was a back door. Struggling, I lifted it up and hurled it downstairs through a back door and onto the ground, where I watched it die. Then it arose, leaving a lifeless wraith on the ground.

I confronted the ghost with a bold, powerful voice: "Rise up! Rise up your soul and join the entity of God!"

In the distance I heard a man speak. "He was the carrier of light."

Another man said, "I guess he was unable to hold it."

I then saw many people; they looked in curiosity at the specter's body. They were sad, but could neither speak nor weep.

After that, I saw many others of all sorts, most resembling common people. They appeared to be near death. Slowly, they marched toward the gray apparition but seemed unable to come closer than about twenty feet from the specter. They wanted to occupy the body of the ghost but some invisible barrier stopped them.

While these scenes were in fact invisible, my mind's eye interpreted everything as visible. The voice in my mind directed their movements. The words had a power and authority to direct the soul of man. This was all visible to me, yet I seemed to be invisible with a power to direct those apparitions on command. I was in command of my own will; my intuition led me to a powerful and triumphant victory.

Again, I found myself in the future. I adjusted myself in my chair and flipped the desk calendar to Wednesday, June 25, 1989. I was listening to one of a series of tapes that my daughter, Zephera, and I had decided to purchase after a personal conversation with Dr. Stranges.

"Ladies and gentlemen, my dear listeners, you are listening to a series of tapes, Truth tapes. This is Dr. Frank Stranges speaking. This particular tape that you are about to hear will perhaps present some tremendous revelations that you have never heard in your own home church. Jesus has made us kings and priests. We shall reign on this planet, Earth. *You shall judge all of those who come up in the second resurrection*, Revelation 20:4, Daniel 7:22.

Cut-away illustration of the Hollow Earth, by Michael Abbey

"M.R. Gardner based his work on the Bible and on the books entitled *The Phantom of the Poles* by William Reed, and *Journey to the Earth's Interior*, Bear in mind that this planet, Earth, as it stands now, is a shell about 800 miles thick. The interior of this planet is an empty space 7,000 miles across in diameter. There are now two major openings into this planet where the North Pole and the South Pole are supposed to be. The openings are approximately 1,400 miles across or 4,200 miles in circumference. Bear in mind again, that before the great flood in the days of Noah…"

I found my attention drifting as a memory of seven years earlier flashed through my mind. It was September, 1972, nearly midnight.

I was comfortably asleep when I heard a peculiar sound that began to vibrate my body. I arose from my body, stood up, and began to walk. Leaving my body behind, I walked through the wall of my house. I entered the outside world and strolled along the street, observing several people as they passed by. Leaves covered the ground with various shades of green, gold, red, and orange—a perfect autumn setting. I could see a faint blue sky in the background, and the air was cool and crisp. Gentle winds swirled the multi-colored leaves into vortices. I found a park bench, sat for a few moments, and then closed my eyes. I saw a vision of myself dressed in a warm and comfortable grayish-blue suit with a hat to match. I was sixty-eight years old. I had moved into the future in a linear fashion.

In an instant, I disappeared from the park bench. It was immediate—my physical form had vanished. I was not aware of the distance or the time I had traveled; my arrival seemed to be instantaneous. Suddenly, I was with a beautiful woman with dark hair who was showing me a highly unusual structure with many rooms.

The rooms were purple and white. The articles in the room were of different geometrical patterns and shapes. The woman explained to me the location of what she called a crystal accumulator. Pure magnetic energy flowed in and out. She began to describe it to me excitedly, saying that I had been teleported by way of the crystal from Earth to the planet Venus.

I recalled Dr. Stranges' voice on a cassette tape continuing to play subliminally in my subconscious. "All of the other nine planets revolve around the sun. But they are not revolving in one direction. The Earth is traveling in two directions. It's spinning while it's going around the sun. Now, for that reason, because the Earth is constantly turning, revolving, there is no spot inside the Earth that can be called bottom. Therefore, the great, deep space—7,000 miles across—is called the bottomless pit, where Lucifer shall be chained. The bottomless pit has no bottom because it is continually moving."

All the while, my planetary experience of Venus continued as I listened to Dr. Stranges' tape. Without saying a word, I observed and listened intently as I followed the woman through the dreamlike presence of another world through this enormous crystalline structure.

The rooms were arranged in a geometric layout, which activated the cause and manifestation of an absolute universal energy. I was shown all the rooms and briefly told about each one. The crystal was huge, located partially underground and partially exposed on the planet's surface. The rooms shone with light, according to the amount of energy emitting from the crystal. I was instructed to stand beneath this slowly rotating crystal. I could hear the deep cyclic motion and a low-pitched sound. The crystal was composed of purple and white facets rotating in their own circle. It was enormous; I felt as though I was standing beneath a huge planetarium. As I stood beneath the crystal, I was teleported from Venus back to Earth.

Suddenly, I returned to the park bench where I had been sitting. I saw myself stand up and walk against the gentle breeze of the autumn wind. I walked back through the wall and gradually re-entered my body. Once again, I felt the warmth and comfort of my bed and the blankets around me. The sound began to fade as I adjusted to my familiar surroundings. Later, I was awakened by a long, high-pitched pulsating sound that long echoed in my ears and remains in my memory.

Several more dreams and memories were linked to this mysterious crystal. Although I was unaware of it, for some time, Fred had been studying the formation and structure of a crystal power source. Often we have had harmonious and simultaneous thoughts regarding the principles and construction of the crystal. He left me his writings and some math formulas relating to its construction. My father and I shared some of the theory on the mechanics of its construction. He had given a lot of thought to the theories of perpetual motion and had related his understanding to what he termed "from the upstairs." Somehow, the fragments of its memory remain permanently etched in our minds. Fred had demonstrated its movement by using his two hands to energize its spinning movement.

"For the past 6,000 years, Satan has been king over the angels…" Dr. Stranges' insistent voice on the tape overlaps and interweaves throughout my other memories—other dreams. One such occurred at the time of President Richard M. Nixon's re-election in November 1972.

President Nixon appeared to me. I saw him clearly in my dream. He was speaking from behind a podium. He had a large protruding eye in the center of his forehead. It appeared to be strapped on—a piece of plastic resembling an eye.

Turning to me he said, "I know who you are."

I replied, "It doesn't matter who I am, but it matters what I am going to do!" I began to challenge his great authority and power by calling him "big-eye" several times in a loud, angry voice that finally became hoarse. He then countered by reading several names to me, but only one remained in my mind after the experience. That name was Paul Winkle. The next day I found his name in the library in Who's Who in America. He was the head of the Central Democratic Committee. Why had I never researched his involvement in Watergate break-in?

> *For the past 6,000 years, Satan has been the king over the angels in the center of this planet. He is still considered the Prince of the Powers of the Air. In Revelation 19:11, it says he is the God of this present world order. The four men that I mentioned at the convention last evening who are in control of this government—those four men are ruling this government. I call it the invisible government. These men are not self-motivated! They are moved! Manipulated, commanded, may I even use the word "possessed" by the evil powers, because if they were possessed of good powers, of the power of God, the power of Light, this world would not be in the condition that it is today!*

Just after the New Year on January 23, 1973, the ground in Iceland split apart by an earthquake. The earthquake was located in the Westmann Islands in Iceland. The ground opened up—two miles wide and four miles long. The eruption opened a mile-long gash at the edge of the town of Vastminwaeyjar, Iceland.

During this time, and while residing in Watsonville, California, I had another dream about Iceland. Throughout the whole dream I was observing myself. I found myself looking directly into the face of a volcano and reaching deep into its interior. I cupped my hands together and collected from the base a brilliant white liquid-light substance. As I collected this liquid substance, my head and body melted as the fire and tremendous heat blasted high into the air.

I was overwhelmed by the absolute power of this magnificent brilliant white substance. It was the raw, uncontained source of all power. Unafraid in my excitement, I ran down from the volcano. The substance remained and glowed brilliantly upon my hands. Then I saw myself passing through the lives of several different Icelandic people. First, I was the discoverer of the substance, then a child running alongside a white picket fence, and then I was a man walking. Many people and scenes passed before me. Each person I saw was carrying in their hands this same beautiful white glowing liquid substance. They all seemed to be moving in one forward motion.

The Icelandic people were carrying the light and the mystery of its origin. They were all carrying the secret word of Creation. Their oral history went back to the dawn of mankind—the Creation and the Sacred Land, Paradise, Shangri-La, Garden of Eden, Aghartha—the perfect dwelling place of mankind during his Golden Age before the great fall, where the highest ideals of humanity are enacted and peace reigns supreme over all. All have carried the same light in different forms, from the peoples of Europe, Asia Minor, China, India, Africa, Iceland, Greenland, and the Americas. All speak of secrets made known only unto the worthy, the pure, and the innocent. We may all walk the path. We may all bear this Sacred Word to our brothers, if only we can grasp, hold onto, and humbly use the key. Was this the key to Paradise and to this sacred place? I now held this glowing substance in my palms.

As my dream-self entered my house, the glowing substance remained as brilliant as when I had first found it in the volcano. In my dream, I woke Katrina while she was sleeping. I wanted to give this liquid white-light substance to her. When she awoke, the liquid white light instantly turned into stone. I was shocked that I had nothing more than a cold stone clutched in my hands.

Suddenly I awoke. I went into the front room and turned on the television. The news was on. I stood in front of it, staring but not really seeing. I backed away and sat down on the couch. Suddenly my attention was drawn to the television. The newscaster was describing a volcano erupting in Iceland! The people were preparing for evacuation.

Later, I referenced the event with color photographs of the eruption in the National Geographic Magazine, Volume 144, No. 1, July 1973.

During this time in our marriage, it was extremely difficult for me to explain anything to Katrina concerning these experiences. Perhaps, needless to say, my bizarre behavior was taking a toll on our marriage. After the newscast, I went back to bed to tell Katrina about what had happened. During my attempt to explain, her interest dwindled as she rolled away from me with a brief "good night." My feeble attempt to pique her interest failed as she silently drifted from me. Any hope I had of conversation evaporated. Motionless, I remained staring at the ceiling with a great sense of despair. I said nothing further and fell asleep with Iceland on my mind.

I cleared my throat, looked at my mirror reflection, and continued to rehearse the orientation lecture:

"As far back as time itself, legends exist of a Hollow Earth, peopled with a superior race, advanced thousands of years. The entrances to this interior civilization have always been thought to be at the extreme north and south of our planet Earth. Well, I propose to go and see.

"In Dr. Raymond Bernard's book, *Flying Saucers from Earth's Interior*, he speaks of legends throughout the ages about ancient Greeks, mysteries of Delphos and Eleusis, the Elysian Fields of Mt. Olympus. During Vedic times, these legends referred to mountains called by various names, "Tatnasanu," peak of the precious stone, "Hemadri," mountain of gold, and "Mt. Meru," home of the gods. Archetypically, the sacred mountain has its peak in the sky, its middle in the Earth, and its base or origin, in the underworld.

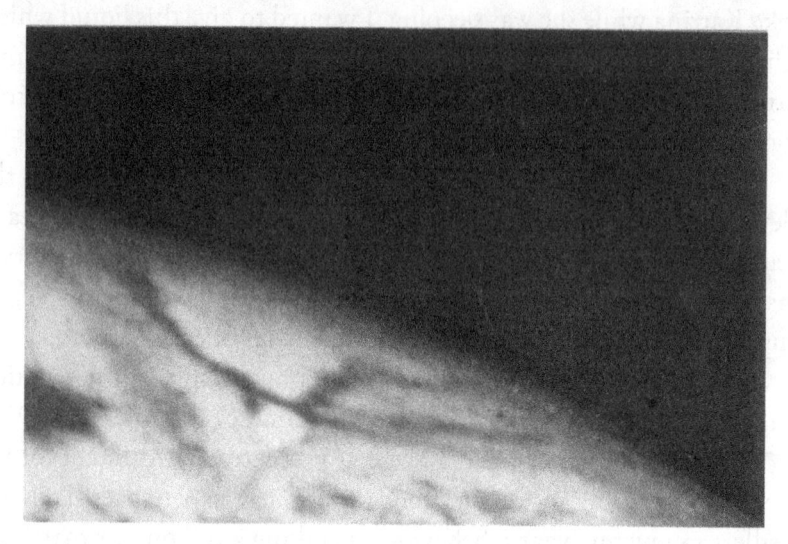

Notice the odd polar depression?

"Shamballah was the Scandinavian celestial city in the land of Asar. Amenti was the land in ancient Egypt, Alberdi in Persia, Tula in Mexico, Chavin in China, or Kalki, sought by the Argonauts in search of the Golden Fleece. All refer to a glorious underground Paradise—in the very center of the Earth itself—where war, sickness, hunger, and even death are but a distant memory. This is Valhalla, Mt. Salvat, Utopia or Shangri-La. Even within the beliefs of the Church of Jesus Christ of Latter Day Saints, the Mormon authorities believed in a Hollow Earth. They referred to it as the Heavenly Father and believed the Lost Ten Tribes of Israel entered into a place beyond the North Pole. This is the Garden of Eden that, together, we will most surely find.

"From Washington, DC, our point of departure, we will fly to the capital of Reykjavik, Iceland. Reykjavik is the metropolis of the far North. It is a country where men are filled with an abundance of compassion, and the women have a reputation for their beauty, inner grace and serenity. I have been there many times myself and I can vouch for this.

From Out of the Depths 189

"This is a country of Viking descent, a beautiful rugged country that is rich with folklore and ancient legends of the Ultima Thule. From this most heavenly earthly launching post, we shall then venture forth into the unknown land—that wondrous "land beyond the pole."

The Aurora Borealis is a northern light phenomena that hasn't been completely expained by science.

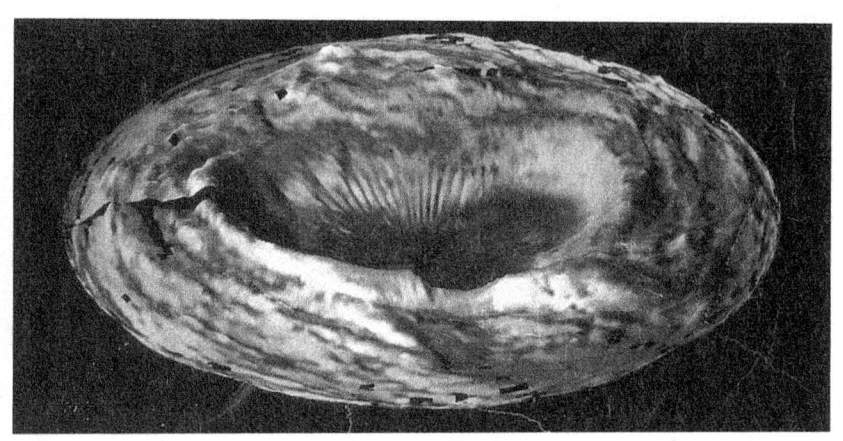

NASA reports the ozone hole visible above Antarctica is a seasonal occurrence caused by the use of CFCS (Chlorofluerocarbons) from spray cans.

Chapter Twelve

THE DREAM

Today's dream is tomorrow's reality.
—William Shakespeare

I FELT A MAGNETIC pull directly over my heart and chest. I was dreaming. I was in Montana and it was May, 1973. "Something is trying to contact me," I said to Katrina. She was still asleep. I lay quietly on my back, trying to relax, but the summons returned. My whole body floated gently and silently upward; I felt a sense of peace within a vast dark space. I was visible only to myself and I moved within an awesome, limitless space, perhaps like a child before birth. I floated upon the waters of God's creative life. I pleaded for help to complete the unfinished man. I became filled with knowledge to complete the fulfillment of God's great plan for man. Yet I still felt fragmented. I was compelled, driven to unite the fragmented pieces of God's great plan that are scattered among humankind. By the power of pure being, will, and prayer, I saw humankind as a single image of God. I was immersed in a bath of emotional energy. It began to magnify. I put my fingertips to my forehead. I was caught by the grace of God's time. Soon the awakening of this mystical center of the third eye would happen in full power. I begin to summon the cosmic will of our Creator.

The phone rang, snapping me out of my time trip to 1973. "Why couldn't she understand?" I muttered to myself as I reached over the bed to pick up the handset. "We might have still been together. God, Katrina. Why couldn't you understand?"

"It's not Katrina, Dan. It's Wanda." The tinny voice reminded me. "Wanda?"

"Wanda. Remember me, your short-tempered, next-door neighbor?

Where've you been? What've you been thinking, anyway?"

"Oh, Wanda, yeah, I'm sorry. I was just remembering some weird experience I had in Montana, and trying to get through to Katrina about everything that was going on with me. If only she could have shared it all with me."

"Danny-boy, you're talking impossible stuff! Jesus, you tried ten years with that woman!"

"I know. I know. So much was going on. No wonder she thought I was crazy. At least I had Steve Singer, my childhood school friend, to talk with. What a kick that guy was. Didn't I tell you about the UFO thing we saw that year?"

"I don't think so—not that one, anyway."

"Who was it? Oh yeah, I remember now. I told it all to Diane when she was transcribing those Ritter von X tapes for me. She ended up walking out on the project not willing to be a team researcher, just wanted to do the research on her own. You know about that, right? But that's who I told, I remember now."

"So, tell me, too."

"It was up there in Montana, that same year. About eleven o'clock one night, I stood outside our small two-room cabin where the family lived. I was talking to Steve, wrapped in a clear Montana "big sky" starry night, when we noticed with curiosity several circular orange lights moving south across the sky. While we observed, to our amazement we noticed a light suddenly flash above us. It was like a flash of lightning that blazed from the sky, strangely influencing us. We looked at each other and knew we were somehow strangely influenced. Then we parted, knowing something peculiar had just happened, but without giving it any real attention.

"Katrina's two children from her previous marriage, Lesa and Art, were sleeping when I entered the room. So as not to awaken them, I silently got into bed and lay upon my back. I seemed to have fallen into what felt like a hypnotic sleep. I don't know how much time elapsed before I noticed a tall, gray being, shrouded in shadows, walk through the wall. I lay still, cognizant of its movements as it approached me. Gently, with its long outstretched hands, it turned my head to the left and inserted a small object just below the base of my neck. I felt no pain and drifted to sleep as the being departed through the wall.

"The next day, the Christian preacher, Jack, came over to invite us to an evening gathering. He was our landlord, as well. Katrina and I agreed we would meet him at the local church. I didn't fully recall the experience I had had the previous night until later that evening. The children went out to play while I set up a meditation sanctuary in the bedroom. Upon a desk with a mirror propped against the wall, I arranged three Egyptian pillar-shaped candleholders with an incense burner. I lit the candles and the rose-scented incense, and invited Katrina to participate. I began to make vowel sounds that we had been familiar with during our times in the Rosicrucian Supreme Temple meditations. I called to Katrina a second time but heard no reply. I continued my meditation, but I wanted her to be with me so that we might enter the next day in a moment of peace.

"Pulled from my mediation, I got up and entered the room where Katrina was watching TV. When I sat next to her and asked why she didn't want to join me, she said I was under the influence of the devil with all these things and had no real understanding with God. I was deeply saddened and hurt by her comment. I asked her if this had anything to do with the preacher. She said because these things were not of the Bible, they were anti-Christian. Distraught, I could not hold back tears. I returned to the room where I had left the candles ablaze and incense burning. I laid my head upon the desk in solemn silence and felt utterly alone and abandoned by my wife.

When she entered the room, she told me she had indeed been talking to the preacher, who told her that we must abandon these satanic occult things. Furious, I jumped up and spoke in an angry voice, "Should I blow up the pyramids, too?"

I yelled in her face; she threw everything off the table that I had prepared for meditation. After smashing everything, she stormed into the kitchen. I grabbed some of her things and threw them against the wall. We began damning each other in verbal battle. I grabbed her, holding her by the shoulders, desperately trying to yell sense at her at the top of my lungs. With her screaming at me to let go, we careened angrily around the room. She grabbed my beard with both hands and pulled with all her might. She was like a revenging Indian spirit that wanted to kill me, and before I let go, I was going to be scalped by her bare hands. I grabbed a deer rifle that the preacher had been trying to sell me, resenting him fiercely. I intended to throw it out into the street. I managed to get the front door open. But suddenly we were struggling in anger again as we wrestled ourselves to the floor with the rifle now in both our hands.

I was still yelling when I fell—trying to surrender to a silent inner compassion and emotional exhaustion. I crawled to the corner near the front door. Just then, someone came to the door to see what the noise was about and Katrina screamed, "Call the police, quick!" Next, she grabbed my step-daughter's jump rope, hysterically screaming that I deserved to be punished. She whipped me several times while I turned my face to the wall. I sat in total submission while she whipped and bruised my back.

By this time, the small neighborhood had been alerted to our wild domestic battle. A number of people arrived along with two sheriff deputies. The preacher and Katrina and the others watched me being handcuffed and taken away. I spent the night in jail.

The preacher came to visit me the next day and asked if I would pray with him for forgiveness and salvation. I consented. After he left, the deputy came to take my picture. Walking toward the camera, he said, "You didn't have to let that preacher in, and besides," he added, "I would never let him do that to me."

I told him it did not really matter. He asked me to remove my shirt. He noticed the large long welts on my back and asked how in the hell had that happened. I explained it had happened during our argument, that I just sat in the corner letting her whip me several times. He called to the other officers. They looked at my bruises and welts and asked, "Your wife did this to you?"

Another added, "I would never allow any bitch to beat me in this way. This will be thrown out of court, for sure."

He asked if I wanted to press charges against her, but I declined and thanked them for acknowledging me. Katrina was outside waiting for me; apparently she had come with the preacher. The officer escorted me back to jail, saying that this case was over and I would be released later that afternoon. I agreed when he asked if I wanted to see Katrina while I waited in jail. Katrina came up close to the jail door. When we touched hands through the open bars, she reminded me of a vengeful beast that was now subdued and repentant.

She slid the Rosicrucian manual between the bars to me. It was her way of saying, "I have forgotten, can we start again, please?" Our forgiveness of one another was genuine, I thought. We later had a quiet dinner back at home and began to repair our damaged hearts.

Throughout the following month, I was in a state of physical and emotional exhaustion and I slept much of the time. One hot June afternoon, I was asleep and face down on my bed. I suddenly accelerated out of my body with a rushing noise. Once again out of my body, I saw a movie with light-filled images move rapidly past my eyes. When the acceleration ceased, I was floating up through my back. I could see my physical structure, my head and hair, and then I was completely free from my body, although I was still conscious of my body lying there on the bed. My entire psychic essence was pulsing with a white, starry, flickering light.

I drifted to the doorway and stopped there. I turned and could still see myself as a visible being, lying on the bed; I was positive it was me and that I was clearly visible. Then I floated into the kitchen where Katrina stood. I told her, "Within this world we love you." She did not hear me directly, but intuitively she received my message as a warm, loving thought that would fill her life with joy and peace and harmony.

I returned to my body in the bedroom and entered it as though donning flesh garments. I seemed to have poured into my body. First, I slowly put my hands in. Then my arms sank silently into their place of flesh. I felt the radiant light of my psychic body settle easily into my legs. I began to move my hands, and then I turned my body. My shoulders had sunk silently into their proper physical places. I saw the hair on the back of my head as I floated into my body. Suddenly, I awoke and regained my faculties.

"Diane," I said, the receiver still to my ear.

"What about Diane?" Wanda asked.

"It was Diane. That's who I told that story to. Diane. That was in 1984, before she walked off with other notes and tapes from Ritter von X."

Still on the phone with Wanda, but not talking, I recalled the first time I had told the same story to Diane. Her image flashes before me again. She was holding her eyeglasses clutched between her teeth, listening intently, and taking occasional notes on her steno pad.

"Then I remember moving my body around, sitting up, then standing and walking near the bedroom door. As I observed myself there on the bed, I was completely rested and at peace." I remember sipping fresh rose hip tea while answering Diane's questions.

"And, during all of this, were you also thinking about Iceland, or the Hollow Earth?" Diane asked, trying to get a perspective on how the cosmic and metaphysical journey related to my current research.

"No, well, yes. I mean, of course it had come up in dreams, like the one with Art, but so far, there had been no contact with anyone, or no books that sparked any interest in going anywhere like Iceland."

Diane punched the stop button on the cassette as I continued, "The center of the Earth thing never really occurred to me until last year. It was always there. I mean, my images and the out-of-body stuff, and now that I think about it, everything! It's always been there. Only, I wasn't fully aware of where it was all heading."

"When did you really know it?"

"It all started to come together that year, after the divorce—not until late in '81. That's when I started to find these books by authors like Raymond Bernard and Lyon. Then I called you in on it, after my first contact with—" I stopped for a moment.

"You mean with Ritter von X?"

Our eyes made contact for a brief second before I carried on. "Yeah, so that's when I made plans to go to Iceland."

"Let's get back to Montana." Diane reached for the cassette to record again.

"Where were we?"

"You were just getting back into your body."

"Oh yeah, right, now I remember. These became regular, almost weekly occurrences. Once, I awoke in a sphere of deep peace that was like an early evening dusk with a hint of vaporous gold. I heard a voice calling my name, a voice that sounded like my mother's. She called my name repeatedly. To make certain the voice was my mother's, I answered, 'What?' several times. When I was finally sure it was indeed my mother's voice, she said, 'Read the book, *Sepher Yezirah*.'

" 'What?' I asked, to make sure I was hearing her correctly.

"Read the book *Sepher Yezirah*," she repeated, distinctly. Then the sphere soared away from me, leaving me in my waking consciousness.

"I began to think about why physical objects such as the planets and some metaphysical experiences appear to be round or sphere-shaped. I reasoned that when physical and metaphysical observances and experiences occur, the physical eye (being round) projects this deception because so-called reality is the product of individual reception and assimilation of vibrations received. Reality as we know it through our vision is the law and order of consciousness and perception. Actuality, on the other hand, is the law and order of vibrations. The truth of the experience is on the vibratory level, appearing spherical so as to be seen by the human eye. Planetary spheres are an illusion by sight. The spheres, then, are simply vehicles of expression, not the expression itself." I paused for a moment to gather my thoughts.

Diane sighed, "Danny, it's getting late. If I'm going to be able to maintain this, I've got to get some sleep. Let's go on with it tomorrow morning." Diane was clearly exhausted. Her voice had a pleading tone, on the edge of becoming a genuine whine, as she punched the stop button on the recorder, gathered her things, and headed for the door.

"I understand." I followed her to the door. "See you tomorrow."

Unable to call it a night, I returned to the dining room table and resumed reading a dog-eared yellow paperback book entitled, *Flying Saucers from the Earth's Interior*, by Dr. Raymond Bernard. "What if I could get them all here?" I mused aloud. "I wonder what they would all have to say to each other."

The mist of my imagination began to rise from the dining room floor, becoming a swirl of dense fog over the books on the table: *The Hollow Earth* and *Flying Saucers from Earth's Interior*, Dr. Raymond Bernard; *Phantom of the Poles*, William Reed, 1906; *The Smokey God*, ghosted by Willis George Emerson in 1912, for Olaf Jensen and his father; *The Poles, or Have They Really Been Discovered?*, Marshall B. Gardner, 1896. Out of this thick fog of my imagination, the would-be voyagers emerged and gathered silently, seating themselves with me at the dining room table.

"Would anyone care for coffee or tea?" I was cordial and unaffected.

"Organic, I presume?" Dr. Raymond Bernard took up the offer. He was a tall, bearded man in his late fifties, weighing about one-hundred-and-sixty pounds. His attire consisted of a loose, white embroidered Brazilian shirt and white jungle pants.

"What? Sure, of course, sir." I remembered Dr. Bernard's studies concerning food being poisoned by chemical sprays and artificial fertilizers in current mass commercial farming industries. "Rose hips, from right out here on the deck. I boiled them down a couple of minutes ago."

"Marvelous, my good man!" proclaimed Dr. Bernard, with an eloquent sweep of his hand.

"What happened to you, anyway?" I asked boldly. "You just disappeared. Did something in a subterranean cave in South America try to eat you alive?" I smiled, hoping for a real answer from the doctor. But he replied, "That, my good man, shall have to remain a mystery."

"Let us dispense with the niceties, gentlemen." The new voice came from Marshall Gardner, dressed in a dark brown double-breasted pinstriped suit. His vest showed gaps between the buttons, a vain attempt to cover a gibbous beer belly. He seemed a little intolerant of his compatriots' parlor room manners. "Your tendency to discount the seriousness of this gathering could well sabotage our entire purpose here. It is my most humble suggestion that we do get on with business. Tea, if I may say so, can be had once this matter is settled." He sat down resolutely, lighting up his pipe and crossing his legs.

I pointed to the sign on my front door: Thank You for Not Smoking. "No offense intended, Mr. Gardner. But—"

"Of course. Well, now where shall we begin?"

"We have already begun, Marshall. My contention is that we are right on the very edge of the end." This speaker was William Reed, a wiry man of incalculable age. His mustache obscured his upper lip. His heavy jowls, piercing blue eyes, and thick, curly blonde hair marked an individual used to having his say. "And, that it is supremely urgent we aid this adventuresome young man in the most expeditious fashion. I, for one, suggest that he organizes a personal expedition to the center of the Earth, at the earliest possible moment."

"Whoa," I interjected, getting caught up in my imaginary visitors. "That sounds like a great idea to me! I'm already working on the orientation lecture. I'll get a few people together, hire a plane, and we'll settle this thing once and for all!"

A voice of reason rang in with a caveat. "Daniel, my good man, we are all quite aware of your enthusiasm, as evidenced by extensive research you have already done; you are well aware of the theories and the historical documentation of previous journeys to this subterranean Land of Paradise. You know of the race of tall, beautiful dwellers that live up to one thousand years and more in a highly advanced society. You probably also know that they are the creators of the flying saucers powered by electromagnetism from the atmosphere—we have been witnessing these UFOs since Biblical times. You may also have read about the gigantic mammoth creatures, one of which was actually observed during an expedition by Admiral Richard E. Byrd—"

Marshall Gardner suddenly cut off Dr. Bernard in mid-thought. "These are not prehistoric animals, mind you. They originate from the Earth's interior. Those found frozen in the ice had been carried to the surface by interior rivers, and they simply become stranded when they ventured too near the icy polar orifice." Gardner seemed obsessed with this particular detail. "Those are not prehistoric mammoths that have been found! It is very important that you understand that, Dan. They dwell at this very moment"—he pointed to the living room floor—"right here under your feet. Then, alas, the poor creatures probably fall into crevasses in the ice, where they are trapped. Unable to escape, they suffer an icy demise and are later discovered somewhere in Siberia." Gardner's long sigh was followed by silence.

"Yes, well, Danny," William Reed finally broke the silence with a note of encouragement, "We all know you have done your homework. That's why we are here to prepare you for an actual journey."

"Here, here," the experts agreed.

"For one thing, as you know, I believe that the red, green, and yellow-colored snow at the Polar Regions is due to pollen. Now then,

Danny," Reed continued, "the black, of course, is soot from volcanoes, which must have all come from the Earth's interior. It is the only possible source. Specifically, as you have already learned, the crust of our Earth is a mere eight hundred miles in thickness. Let me explain something to you about gravity. The same gravity that pulled objects on the outside of the sphere inward would thrust objects inside the globe outward. Do you see what I mean? Therefore, a voyager could sail directly over the edge of the polar opening, blithely as you please, without ever being the slightest bit aware of where he actually was, do you see? Of course, the question is, why? This is perfectly simple because of the opposing pulls of gravity, and the ever-so-gradual declining curvature of the entrance. Furthermore, I venture to say that all, or nearly all, of the explorers have spent much of their time past the turning point, and have had a look at the interior of the Earth, whether they knew it or not!"

"Something like a God," I mused.

"What's that, my good man?" Dr. Bernard questioned.

"Oh, never mind." My voice was nearly inaudible—I was speaking to myself. "I just said that's just like God. I mean, whether you believe in God or not, His Divine order—His ruling force is influencing our lives, whether we know it or not. Gravity, simple, yeah, I understand."

"One other essential detail," Reed continued, unconcerned with the side conversation between Bernard and me, "deals with the auroras and the source of light for the interior world beings."

"Please! Mr. Reed! Not this again, Mr. Reed!" William Gardner's intolerant anger flared. "You continually insist on maintaining the most absurd notions about this whole affair! Our outer sun does not light the Earth's interior. For the last time, please! Remember where we are. Must we continue this unrelenting discourse throughout eternity?" William Gardner looked disgusted as he left the room to light his pipe out on the deck.

"They get their light, Danny boy," Reed continued, "from their own sun; you can trust me on that. The northern lights are merely reflections from the ongoing interior prairie fires and continually active volcanoes."

"It sounds like some sort of hell to me," I muttered under my breath. I was the one who had invited these guys here and now they were taking over. It seemed like the best time to offer my guests some tea. I filled a tray with cups and poured. "Let's take a little break, here. What do you say?" I looked up to find they had all disappeared except for Willis George Emerson.

"I'll go along with that. Do you want to hear about Olaf Jensen?"

"Who?" I sipped at my tea, finding it far too hot. "Who's Olaf Jensen, Mr. Emerson?"

"You have it right there," he said, pointing to The Smokey God. I ghosted it for him in 1908. We were in Los Angeles at the time. Olaf had been confined to a mental institution since he returned from a two-year stay in the Earth's interior. Actually, he'd gone with his father, but the older man never returned. He died under an iceberg when their sloop was crushed, but a Scottish whaler rescued Olaf. Olaf tried to tell his story, but they just locked him up and threw away the key."

"The Smokey God, yeah, I remember reading this one a few years ago." The small volume felt light in my hand as I flipped through the pages. I was surprised I had forgotten its author.

"It's a story all in itself. I'll tell it to you if you want me to." Emerson looked to be around my own age. Sincerity rang in his voice as he continued, "The story has already been told—actually. Jules Verne's tale became a movie."

"Yeah, I remember. A good one too—I'll never forget it. Geez, wish I could get the well-known 1950s Hollywood actor, James Mason, to go with me." I could visualize my own journey to the center of the Earth.

"Well, perhaps it's better if you reread it for yourself. Just let me say that Olaf substantiates many of Reed and Gardner's claims such as the magnetic irregularities at the poles, wind-blown colored pollen, mammoth bones, that kind of thing. It makes a credible story."

I stifled a yawn. "Sorry. I'm not being impolite. It's just that I haven't gotten much sleep in the past two months." As I closed the book and sipped my tea, Willis George Emerson's image faded. I realized the evening was definitely over for me.

The next morning Diane arrived at seven o'clock sharp.

"Hi, right on time." I opened the door for her. She stalked past me, sat down, and turned on the tape.

"Let's go." Her first words were abrupt. "Let's get back to Hungry Horse, Montana."

I continued telling my story.

It was 1973. I had received several books by mail, one of which I had planned to give to my landlord, the Christian preacher. My friend, Steve and I went to visit him. As I had promised, I presented him with this book entitled *The Mystical Life of Jesus Christ*, by Ralph M. Lewis.

He invited us into his house and we sat comfortably on his couch. Without any gesture of thanks, he began to thumb through the book, and then called excitedly to his wife. To us he said, "This book and all its information is a blasphemy against God and Jesus Christ."

Puzzled, I looked at Steve. He, too, was totally taken by surprise. "I assure you that this information is not intended to cause any sort of harm," I said.

"Why is the information blasphemous?" asked Steve.

The preacher exclaimed with a growing anger, his voice rising, "Any writings other than the word of God found in the Bible concerning Jesus Christ would originate from the devil."

His diatribe continued. He attributed this pronouncement to a dream he'd had several months ago about a particular kind of demon that sat upon a wooden fence and sneered at him with vengeance and evil intent. He said we reminded him of that kind of demon who dares to challenge the word of God.

Handing the book back to me, he said, "I'm sorry, but we cannot accept this mystical crap."

His wife sounded almost hysterical. "They are from the devil," she ranted.

Steve reacted. "Are you not challenging the Lord God right now by making a baseless judgment that we come from the devil? It sounds to me as if you are putting other gods before yourself. So who do you say is being blasphemous?"

The preacher turned to his wife, "Now they are speaking as gods!"

"Gods?" exclaimed Steve. "Of course—we are gods, are we not?" Steve and I looked at each other and laughed. It was probably a good time to leave.

The preacher began moaning the name of Jesus, sounding like a raging ghost caged in a dark dungeon. "Now they believe they are gods," he repeated. Reaching behind his piano, he handed me a long coiled phone cord, saying, "If you believe you are gods, and then turn this cord into a snake."

I took the cord from his hand and gave it to Steve. I told Steve to show him that God is God! With a strange sense of humor and a little theatrics, Steve walked toward the preacher while shaking the phone cord near his face. "Why do you tempt the Lord God right now?" Steve hissed. "Because you know that we cannot do this? Then we will and we must!"

Taking the drama a little further, Steve handed the cord to me. I tossed it in the middle of the front room floor. Steve ran over and bent over the cord. We were both chuckling. Steve jiggled his hands over the phone cord and said in a deep, low rhythmic voice, "Boogie! Boogie! *Dominic go friscum!* Boogie! Boogie!"

But the moment of humor was over. We were suddenly annoyed and dismayed, and just wanted to get out of the building. We apologized for our misconduct and hoped that we could discuss the whole matter another time. Closing the door behind us, we left—the phone cord unchanged.

I told Steve, "I really thought for a moment you would turn that phone cord into a snake."

"Yeah, me. Too. How did Moses accomplish it, anyway?"

"Fucked if I know," I said. "Thank God you didn't. Otherwise the preacher would have killed us."

That evening, Art, then four, asked for a book to read. I gave him the Bible and told him to put it under his pillow and sleep on it. The next morning he told me that God had come to him. I grabbed my tape recorder, and this is what Art said:

"In the beginning of the Bible was the Holy Spirit of all spirits. The word of God was the word of Moses. He will come with the Holy Spirit and guide all people. The God of Jesus and Moses will be on this planet. They will guide all spirits and people by the God of the Bible book. He will take care of us and all people and everyone around here. This means our world is hidden behind trees and the Holy Spirit around them. Listen, the Holy Spirit is quiet and walking around here."

I transcribed Art's recorded message and read it to Steve. It was nearing ten o'clock in the morning, so I suggested to Steve and my family that we go to church and share this information with the preacher. Just before the service began, Art and I walked up to the preacher. I told him that I had something special to show him, which my son had told me earlier that morning.

First he read it to himself. Then he exclaimed aloud to his small congregation of about forty people, "The Lord has spoken to this child today." He then read it out loud. The crowd was overwhelmed.

In a rush, they left their seats and approached Art, but the little boy was shy. He slowly crept behind my leg in an attempt to hide. It was obvious he was feeling uneasy, but the crowd had many questions for him. He answered some but wanted me to take over. The misunderstanding between the preacher and me the night before seemed to dissolve, all because of innocent young Art, who had spoken a truth about an experience he had of God.

The following September, we moved back to California.

"Dan!" Wanda whispered. "Dan!" Her voice on the phone suddenly jarred me into the present.

"Yeah, Wanda. What?"

"Well, go ahead."

"Go ahead and what?"

"Go ahead and tell me about the UFO thing with Steve."

"Wanda, weren't you listening?"

"What are you talking about, Dan? You just told me you were going to tell me the story about Steve, and then you just sort of blanked out."

"Maybe later, Wanda. I'm kind of worn out now. I'll call you later, okay?"

"Sure, Dan, fine. Are you okay?" Her last words sounded faint as I hung up the phone. I rolled back over onto the bed and instantly fell to sleep.

Chapter Thirteen

❧ LILY ❧

*A woman's sweetness is the nectar of the Gods
and with her love I could touch the face of God.*

—Danny L. Weiss

While I was sleeping during a forty-day fast one spring morning in May of 1975, a liquid blue-white light poured in through the top of my head and filled my whole body. The light replenished me with a surge of cosmic energy. While I was being saturated by this radiant cosmic force of power, the intense magnitude of the light subsided. I entered a vast bluish atmosphere that was infinite in all directions. I found myself inside the soul-personality of Katrina and visibly conscious of a dim, bluish light. She was pregnant with our daughter, Zephera. While she slept, I could feel her breathing around me. I could also see Zephera, in the form of luminous reddish-orange flames emanating inside of both Katrina and me—the causal beginnings of life. All three of our soul-personalities were one within a bluish-purple atmosphere—one intrinsic soul. Zephera was captured in the bosom of her mother's soul, and I was within their center. In this moment of supreme cosmic awareness, I decreed an affirmation of goodness. "Let health prevail within her, forever." While I intoned, my hands were extended into the shape of a cross. When I had finished speaking, I returned to my normal waking state. I was pleased and thankful that such an experience had taken place.

Zephera was born November 18th of the same year. A few months before her birth, I saw my new daughter in a dream. I saw, with clarity, her cheerful bright face and radiant blue eyes, while Katrina held her in her arms. I had been present when her first breath was taken, and now I saw life pour into her luminous soul. I was astonished at the revelation:

my eyes opened along with hers into the world. For me, it seemed like the first time.

In succeeding days, events between Katrina and me became chaotic. My long water fasts were helping me become aware of a haunting, ancient subconscious karmic pattern. During the fast, I was allowed to observe this age-old karmic pattern that would eventually lead me to overcome it. I discovered that this worldly cyclic pattern was living its destined path within me. Fasting and prayer were the tools that changed it.

Ultimately, our emotional drama and the changes we had experienced made us realize we were no longer emotionally compatible. The change took its toll on my family and hers, causing us to separate during the following midsummer. She and the children moved in with her mother. I took my books and the dog and cat, and I sought out my brother Fred, who was living in Montana. He and I began to discuss the Edgar Cayce writings, in particular the readings concerning the suggestion of the cataclysmic changes in the western United States.

I remained a short while with Fred, and then went on my way to meet Lorraine Richards. Lorraine was a well known psychic and a survivalist with a large number of faithful followers. She was a strong advocate of the Edgar Cayce movement. Cayce was a psychic medium who delivered his messages in a trance state. Lorraine and her group had established a psychic headquarters in Williston, North Dakota, where they were preparing in the event of a future cataclysm. They had sold most of all their personal belongings and had taken only the things they felt necessary to rebuild and replenish a new world.

On my way to North Dakota, I spent the night in Glasgow, Montana. I left my cat in the car with the window rolled down a little for air. My dog slept in the motel on the floor beside my bed. As sleep slowly overcame me, myriad life patterns began to roll before my mind's eye like a movie of archetypical images—images we humans often find ourselves bound to. I knew that the forty-day fast was bringing me closer to significant changes that would free me from many old ways of thinking—old ways of moving through this world, and old ways of seeing things and relating to people, especially women. I saw an archetype of the Harlot, bound to the wheel of life, unable to free herself. She replayed her role, over and over again, drawing in all unconscious males. I had yet to overcome this world karmic pattern. It was my understanding that women were held in bondage, victims, just as much as their male counterparts. Prostitutes—in fact, most women of this planet—are made into slaves, I realized. They are kept in invisible bondage to maintain the power and strength of the false gods of our worldly system and order. Women still remain captives and many have submitted themselves to a false idea of God and their self-image in this world.

When husbands, sons, wives, and daughters return to the sacred triangle of love, light, and life—by means of submitting to the laws of love within them—they shall escape the worldly bondage of despair, pain, and death. We are the temple of the living God and the life source of the world. And the way humankind supports it, suffering is the tragic result of those rare individuals who remain strong and face their adversaries every day and night by the power of the living word. Still, they are made to suffer. The Word is the presence of the living God. The world consciousness is the image of mankind riding upon the world's wheel of time. To remain in ignorance and refuse to be guided by the light and the law of love is setting oneself upon the path of a fate worse than death. That path is the Second Death.

Suddenly, I was wide awake. I knew it was time for me to confront my inner darkness. I no longer could afford to break the covenant of God's law. To overcome is to forgive; to ignore the law is to be judged by the law. It was time to let Katrina know of my karmic past that had been the very core of my destruction for many lifetimes. I got up and went to the telephone. This call was essential for my soul to remain in existence.

"Katrina, I was with a hooker during our marriage." Half-asleep, Katrina began to slowly awaken as I spoke, as I unburdened myself. But it had to be done. To my amazement, Katrina understood and agreed she might have been part of what drove me to it. With a great effort, she began reassuring me with compassion and humor, rather than angrily attacking me as I had expected. She understood the sincerity of my repentance. I thanked her for her forgiveness and explained in detail the particular circumstances that had led to this act.

I realized that by collaborating with the consciousness of prostitution, a part of me was allowing the archetypical image of the world image to be perpetuated. I had actually empowered the prostitute, by the power of my own being. This had to be changed.

At that moment, I realized this false image or entity was alive not just in any woman, but alive inside of me. I was ready to accept the challenge and change. Most importantly, it was a part of me that was allowing her to be a servant to this false image. It was that very bondage that I came to understand and needed to overcome. I could no longer avoid these situations.

Once again, it had been the tools of fasting and prayer that initiated the changes that tore down and destroyed the false perceptions in my mind and heart. Situations not yet known to me, with regard to my own family, would come to pass. Ironically, our marriage and family would be sacrificed because of my choice to destroy this false god, the god of the world that lived within me. Only later would I discover that the whole act would be played upon the world's stage within the hearts of all humankind.

The next morning I went to feed my cat, Black Kitty, but to my dismay, he had squeezed through the car window and escaped somewhere into the town. While searching for him, I met an old man and some kids on bicycles. They said they would love to help me because today was Friday the thirteenth. The man thought it would be a good idea to announce that a new person was in town and had lost a black cat, so they arranged for a local radio station to alert the whole town. Within a couple of hours, my cat was found. A small girl who had seen him playing on her lawn brought him back to me. I was pleased and truly amazed that the whole town had co-operated to do such a favor for a stranger. It seemed to be a good omen, a sign that perhaps I was finally ready to balance an age-old karmic cycle.

I continued my search to find Lorraine Richards; however, it seemed my good fortune was not to last. When I arrived in North Dakota, my generator burned out and the fuel pump froze. I rented a small motel room, left my animals there, and walked approximately three miles to Lorraine's place, a large white house. Heavy rain completely soaked me.

A woman answered the door. While I stood freezing in the dark and the cold rain, she emphatically denied any knowledge of Lorraine or any psychic activity within the house.

Feeling defeated and confused, I walked three miles back to my car in the icy rain. When I arrived back to my motel room, I was delighted to find that my animals were happy to see me, and I was just as glad to see them. The next day my auto repairs were completed and I left Williston immediately—heading back home.

Subsequently, there were minor earthquakes in North Dakota, Colorado, and Wyoming. Tremendous rainstorms flooded many areas of North Dakota and Montana. On my way to Mt. Shasta, California, I began to make my prayer of repentance in God's name. I simply asked God to help me end this cyclic repetition.

I wanted to overcome this mental pattern and be free of its cycle. I wanted to have a new existence with a special woman in my life. I realized that personal karma is equal to world karma. This false god to whom I had been adhering declared—insisted—that the system and the order of humankind had already established. With a great deal of prayer, I began to attune myself to the god of my heart. With an added thrust of my deepest emotion, I knew that my message was being heard. I also knew that my time was near for testing—a test that would bring me face-to-face with a power to overcome this seemingly eternal merry-go-round of man's outer awareness. It was my time to free myself from this repetitious and treacherous cycle. I desired to look deep into a woman's eyes and soul. Through this connection, together we would reveal the plan of human destiny. I would find this only through this imaginary woman, and her embrace of love and wisdom.

I arrived in Mt. Shasta City, hiked up the mountain, and camped for a few days. The pure ice-cold mountain spring water strangely energized me. One day, after scaling the west slope of the mountain, I stood near the mountain's peak, in search of God. I met a black man there. He was in a lotus position reciting a mantra about love, freedom, and wisdom. I sat next to him, bracing myself by the stones and leaning against the steepness of the cliff. I was consumed with a deep desire to know the truth of my being. This man and I embraced each other. Our emotions blended within us as a warm wind. For a moment, our thirst for truth was fulfilled. I told him I loved him, and he was amazingly receptive.

He paused and said, "It is our love of the God that we seek that makes us free to love one another." I agreed by nodding my head with a smile. We spoke to each other while wiping our eyes with joyful releases of emotion. In that moment, I learned that freedom is the house in which we live and die. I needed to hear more than the echo of my own knock. Freedom is the aura of God's presence in whose name we petition for ascension.

In our discussion, we agreed that we must ascend the ladder of lights to our God consciousness. Only then would we know that our meeting in freedom would complete its eternal journey.

What have we come here to do in the temple of freedom? At long last, the question became answerable. This to me is the fulfillment and manifestation of God's cosmic will. Behind the cause of freedom are the issues of justice and equality—equal rights of every race and freedom for every race of man, forever. The cause of freedom shapes the destiny of man's ultimate completion. In this sense, the soul of man is like a blanket that warms the throne of the eternal. But this can only happen when man is kindled with the light and power of freedom.

My animals and I wandered many miles back behind the face of the mountain. I saw a large tattered teepee just below me, alongside a small bubbling spring. It looked as though it had been abandoned for years. My dog—part collie and coyote—raced toward the spring and drank from the icy water. I followed him down the hill and joined him. My cat hopped along as if she were a bouncing rabbit; she sat and watched us gulp down the water. The fresh cool water energized me. My dog ran into the vacant teepee, where he immediately found rest. With his head slightly poking out, he looked content. I felt as though I had walked into a guest-only exclusive resort. The animals curled up together and went to sleep, and soon, so did I. It was very early in the morning, a couple of hours or so before the sun rose. I was awakened by great waves of energy that poured through the top of my head. These waves of energy were powerful pulsations. The energy increased as the vibrating sound pulsated at a higher and higher frequency. With a childlike excitement, I asked myself, "Should I turn into light, now?" But as I spoke nothing more happened.

The following day, a freak snowstorm in the early part of July made me abandon my camp on the mountain. I decided to camp alone on Lake Siskiyou. The weather became very hot again. I was gazing across the lake in a half-meditative state when smoke suddenly arose with a burst of flames behind the trees. A fire had begun to burn wildly. I watched the flames race through dry, vulnerable trees before fire trucks came and subdued the blaze. I packed up my camp and moved my animals and myself into a motel in Mt. Shasta City.

That evening, a strong surge of energy entered my head. A dark apparition suddenly appeared in front of me. Distinctly, I felt a presence. I asked, "Who are you and what do you want?" But there was no reply. The next day I met many people who were members of the St. Germaine Foundation and Keepers of the Flame, which was headed by Elizabeth Clare Prophet. She was a spokesman for the Masters of the Great White Brotherhood. They shared with me numerous ideas, including their belief in the crucial importance of the destiny of America.

Count St. Germaine (18th Century)
St. Germaine has been described as the greatest of the European Rosicrucian adepts. His prodigious knowledge of history and philosophy was often commented upon in his day. In a letter to Frederick II, Voltaire said that, "The Count de St. Germaine is a man who was never born, who will never die, and who knows everything."

As I walked down the main street of Mt. Shasta City, I noticed several people walking through a small gate that opened into a small white house. I heard a voice that said I was invited, please come in. I accepted the invitation, sitting down on the floor with others in a half-circle facing a very old woman.

She said, "Please close the door and let's begin our talk." Her next few words were softly spoken, announcing that the Master St. Germaine was here. She spoke about the vast importance of the destiny of the United States, her personal contacts with ascended masters upon Mt.

Lily 211

Shasta, and most importantly, how to prepare for our ascension. She was one of the original founders of the organization called the "I Am Foundation" of Mt. Shasta.

Later on, early that evening, I went to the Friends of the Mountain restaurant. A very attractive woman entered the room. I tried to avert my stare for a moment, pretending to be distracted, but she caught my eye and asked if she could sit next to me. She introduced herself as Lily. Our conversation flared like the fire I had seen earlier. Her green-eyed, radiant beauty surrounded me like warming flames. I lost track of time, and soon, the restaurant was near closing. She invited me to her house, a small room upstairs in a large, old Victorian mansion. We entered the house quietly so as not to awaken the owners and tenants. She took me by the hands and kissed me and led me upstairs to her room, where she had me sit in a large rocking chair.

Lily seemed entirely unself-conscious as she undressed before me. Then she lit a candle and some incense. She offered me some herb tea, and knelt before me with her hands extended and her head bowed low. She made me feel like a king. She seemed to believe in me in a most unusually humble, reverent way. Her radiant presence caused an old story to flash before me, a story I had recounted to my father when I was four years old. Lily's nude body moved gently through a yoga ritual as I told her the story.

I had asked my father to bring back some stones from the mountains where he had been working. One night, he brought back various rocks and laid them beside my bed. The next morning, I picked up a white stone that resembled marble. I gazed at it, curious about it. On one corner, I discovered a green radiant gem. I rubbed my fingers on this shining green gem several times to make sure it was real and would not disappear.

Later that day, my father returned home. I ran from my room to show him the bright green gem embedded in this white stone. I picked it up and could not find the green gem on it. I looked several times, but it was gone. I knew I had seen and touched it. It was real. My only thought was, where did it go? How could I tell my father that the green gem was once there and now was somehow gone?

Now, gazing into Lily's piercing green eyes, I thought without words, and said, "You are this precious missing gem and you are to be placed rightfully within God's crown of life."

Her eyes reflected a love deep with compassion and purity of heart. She backed away and a silence, warmed by love, fell upon us. She took a flute into her hands and began to play it, while her emerald eyes cut into the center of my heart. It was getting late and I got out of my chair.

"Thank you for a most pleasant evening," I said, heading for the door.

"No," Lily insisted while her fingers silenced my lips. "You may stay the night."

"I'd love to, you know, but, ah—" I fumbled for words. "I seem to be—" How would I say it? "I am bound by a peculiar karmic tie that I must overcome."

"Understood." Lily's gentle smile carried a knowing sadness. She had said more to me with her silence than through words.

I kissed her goodbye. It was difficult for me to leave her magnetic personality—it was so powerful that I nearly succumbed, my breath almost smothering me. I broke away from her goddess-like force and silently tiptoed to the front door. But she left me again almost powerless with a kiss that melted me with a quivering vibration I shall never forget. I left her arms, finally, and strode out the door.

A path was presented to me. I would be destroyed in the Second Death—the destruction and annihilation of the soul. No creature of God can afford to fail such a personal judgment by Him. I had made a personal vow to God's covenant and was not about to break it. The following passages rang in my ears as a grim reminder.

> *Do not be afraid of the things that you are about to suffer. Look! The devil will keep on throwing some of you into prison that you may be fully put to the test, and that you will have tribulation for ten days. Prove yourself faithful even unto death, and I will give you the crown of life. Let the one who has an ear hear what the spirit says to the congregations: He that conquers will by no means be harmed by the Second Death.*
>
> —*Book of Revelation 2:10, 11.*

Happy and holy is anyone having part in the first resurrection; over these the Second Death has no authority, but they will be priests of God and of the Christ, and will rule as kings with him for the thousand years.
—Book of Revelation 20:6.

Look! The tent of God is with mankind, and He will reside with them, and they will be His peoples and God Himself will be with them. And He will wipe out every tear from their eyes, and death will be no more, neither will they mourn or outcry nor pain anymore. The former things have passed away. Anyone thirsting, I will give from the fountain of the water of life free. Anyone conquering will inherit these things, and I shall be his God and he will be my son. But as for the cowards and those without faith and murderers and fornicators and those practicing spirits and idolaters and all the liars, their portion will be in the lake that burns with fire and sulphur. This means the Second Death.
—Book of Revelation 21.

My God, I thought. If I looked back, I would not only turn to a pillar of salt, but my soul would be thrown into a lake of burning sulphur. I took my car and traveled for a mile. Then, with a hard left, I completed a U-turn in the middle of the highway and raced back to the old Victorian mansion. My whole body was quivering from the memory of Lily's kiss. The excitement was overwhelming, and I had some difficulty parking my car. I ran to the side of the house. As I looked up at the top of the mansion, I saw her standing on a balcony in front of the small-paned glass windows. Her long, golden hair was radiant from the soft glow of the candle she was holding. She looked at me as if she was expecting me again. I met her at the front door, she took me by the hand, and we quietly rushed up the stairs. My heart was in a flutter and vibrating with a deep passion. Why had I returned to her? I felt that she was an angelic host who would free my soul and help me find my personal identity towards an unknown destiny in some mysterious way.

"How am I to overcome a passion that was only my reflection of a karmic memory, deeply rooted within my subconscious?" I asked her. I began to trust her judgment, doing exactly as she asked. My heart throbbed to the glint of the candlelight dancing on her face, her neck, her soft, moonlit body. She began to undress, and asked me to do the same.

Blindly, I followed her direction, and together we slipped under the bed covers. My whole body was tingling with fiery vibrations. My attraction was intense, yet I was held like a ripened apple ready to be devoured in the bosom of her deepest purity and love. I was very conscious of the magnetic force she had upon me. I was falling into her silent will. I was being tested. This test I knew was upon me, though I did not know to what degree or what the outcome would be. I did not lose sight of the test and God did not lose sight of us.

Lily whispered in my ear, "Would your wife mind if I kissed you goodnight?"

"I don't think so."

Once again, her kiss shot wild, pulsating electrical vibrations throughout my body. However, the thought of going beyond one lavish kiss left me with thoughts of the Second Death. This was being made very clear to me. We talked for a few minutes, and then Lily snuffed the candle out. Through her bedroom window, the radiant mountain of Mt. Shasta gleamed in the light of the full moon. Without words, I knew Lily understood my devotion to God and His law. We stayed still. Solitude, peace, and quiet filled the room while the soft moonlight was our very breath and showed the pathway beyond the stars.

Mt. Shasta—Photo taken by Zephera early one evening when she met and spoke to a strange, tall man with piercing blue eyes, pale white skin and long white-blond hair.

Lily's warmth was like the softness of sun-drenched velvet. I dozed, and then awakened suddenly to the sound of her gentle breath; otherwise, there was nothing but silence. I was in a warm sweat and yet comforted by the coolness that poured through the open window. I dozed again. A deep sleep fell upon me like a thick, heavy, warm blanket. My mind seemed to be on fire as several dreams of lovemaking came to my consciousness. Then I fell into a relaxed, stilled state of mind.

I think a couple of hours passed. I opened my eyes with all my attention on Lily. Her eyes were open, fixed upon me in a gleaming stare. I felt the force of my passion desperately struggling against her serene stillness. I moved closer until I could feel nothing but her warm body. At that moment, more than anything, we wanted to embrace each other and make love in this sublime stillness. As I went to embrace her waiting passion, I gently slipped out of my body and entered into a golden, vaporous atmosphere. While our bodies lay at rest, our souls merged into one. Her soul-personality and mine merged into an oversoul. Lily's radiant love was an invitation—the entrance into the temple of the God that dwelled in her heart. We found a profound cosmic peace in the merging of our souls. I rose almost instantly, but at the same time, my body became formless.

We slept within the bosom of this godlike purity. She allowed me to rise above the karmic wheel of my past cyclic nature. It was as though I had stepped behind my mortal self. Now I could rebuild myself toward a greater understanding of the will and plan of the true God. It was because of her invitation into her radiant inner temple of love that I was now free to move toward a new dawn of freedom.

She was like Adam's first wife, Lilith, free from any curse of the Tree of Knowledge inviting me into God's cosmic grace, while my wife, Katrina, was like Eve, teaching me the fall from cosmic grace.

Then it was morning. I awakened to see Lily doing her regular yoga and meditation exercises. She apologized, saying that she had an appointment concerning her upcoming journey to Tibet. With her angelic smile, Lily kissed me goodbye. I sprang from her bed and took a deep breath, feeling as though I had been newly born.

Lily left me with a new day and a new life. I felt free and light as though a heavy anchor had been lifted, unburdening my soul and mind. In silence, I thanked her as I prepared to leave. I left a note on her dresser mirror with God's name written in Hebrew. "Thank you, Lily. May God show favor in all that you do." I left Mt. Shasta City, but inside my heart Lily would always be with me.

Arriving in Santa Cruz, I mutely prayed and thanked my God for the most beautiful event a person could share with another in the sight of his or her God. I was most excited to get back to my family, where I knew I had to complete my karmic responsibility. Within me there was an abundance of light; I began to pray aloud with deep love, respect, and adoration for God.

"Father! I miss my family very much. My baby daughter is dearer to me now than ever, for I have been awarded this light of freedom by overcoming an age-old karmic pattern. I am willing to sacrifice it all and hand it all back to You. I would be glad to offer this gift of light in exchange for those who suffer yet desire to be free in your presence.

"This is my personal quest to fulfill all that remains unfinished in Your holy name. I ask You, Father, from my heart, is Your human family yet free from bondage so that Your name can be exalted and glorified within every heart and mind? On what and where can I rest my eyes? My eyes cannot rest until they reflect the glory of Your holy name.

"Father, I do not wish to reunite my family except that we should come to know You as our Creator, salvation, and the promise of Your new covenant with Your holy name. And this is despite the fact that I love them and am anxious to see them more than ever and share my life with them. We must come together to fulfill the purpose of Your holy name. This family should not come together except in Your name, whose great will is yet to come. Let the human family unite by the will of Your unpronounceable name!"

I cleared my head of the entire quest and any questions relating to it. All of this had come about as a result of the breath of a new day. The promise of a new covenant from our God was near to my soul. The presence of the Creator would invite the world into the promise of life eternal.

The light of Lily's love alone allowed me to free myself upon a greater path of initiation. Lily's love would always be a lamp and pathway to the mountaintop to see through the single eye of God's own being.

Chapter Fourteen

❧ God in Our Midst ❧

For as Jonas was three days and three nights in the whale's belly so shall the son of man be three days and three nights in the heart of the earth.

—*Matthew 12:40*

"Danny, you sure tell a great story."

"Well, get comfortable, Wanda, because there's a whole lot more." I felt like Billy Pilgrim in Kurt Vonnegut's book, *Slaughterhouse Five*—coming unstuck in time. I continued telling my story to Wanda over the phone.

I arrived home from Mt. Shasta and was welcomed to begin anew with my family. We reconciled our differences and decided to end our separation. Katrina and I started to look for another home, finally settling in a small town called Freedom, California. To this day, I find that singularly appropriate.

Once again, our family was reunited. I enrolled Lesa and Art at the nearby elementary school and I went back to college to complete my degree in political science. Katrina spent much of her time at home tending to Zephera, who was then one year old.

One day in September of 1976, during a late afternoon nap on the couch, a great surge of light and sound swept through my head with a roar similar to that of a jet aircraft taking off from some nearby runway.

My thoughts whirled with an endless view of diametrically-opposed thoughts. I slipped out of my body and found myself drifting slowly and falling to the floor. But then I floated into a meditative position where I could clearly view these opposing thoughts. I knew my family was confused and still remained somewhat frightened of me.

I could view all my thoughts through a kaleidoscope of mirrors. Each mirror reflection was a different geometrical shape. It was as though I was looking into thousands of magnified geometrical snowflakes. Each flake was geometrically different and perfectly woven into one intricate divine thought pattern. I became absorbed with discovering where this magnificent thought had come from. I realized that each thought creates its own geometric pattern, and that I was a part of each pattern yet still remained isolated in individual fragments. The inner mediation was a vehicle leading me to divine thought. Each geometrical form had color and was musical as it danced within my view. I was able to understand each form independently and know its greater importance. I could see each thought harmoniously forming itself into a single entity. The point of harmony was, in fact, the nucleus within each geometric pattern. Each thought form had its own power stemming from its nucleus. The nucleus of all thought is the golden light, and the nucleus of all matter is the cosmic mind. I believe this is a mystical explanation of meditation and the power that it serves. I was nearing the oneness of both. I understood what the great Greek mathematician, Pythagoras, had said—that, in the beginning, God geometrized.

A few weeks later, a quivering but intense vibration passed slowly over my body. I was lying face down upon my bed. My entire body was permeated with an electronic surge. The vibration emitted a low-pitched sound that lasted for several seconds. I recognized that these strange occurrences of sight and sound were gradually leading me into a greater experience. I was receiving a new realization; I was absorbed with this divine thought pattern.

Once again that month, I dreamed of a flying saucer. First, I saw it soaring across an early morning sky. Suddenly I became the pilot, viewing mountains, trees, lakes, and buildings. The dream suddenly stopped. I found myself raised above my physical body. I lay still and quiet. While in this state of awareness, I decreed, *"In the name of the Almighty presence, please send forth Your healing light."*

Immediately, a spiraling funnel-shaped light energy entered certain parts of my body, penetrating into my flesh, into the bones of my arms and then my legs and pelvis. This spiral light energy was warm and painful by turns because of its intensity. I could hear faint voices and see vague outlines of three light beings. I called them celestial healers.

On another evening during that same month, I awoke to see and hear ghostlike entities whispering into my ear. They seemed gentle and friendly. They spoke in soft feminine voices. They were harmless, yet I tried to avoid them because they were disturbing my sleep.

Those whispering "female" entities reminded me of another time, later on in 1984. I recalled feeling the same uncomfortable disturbance while reading another book, Aghartha, written by Sungma Red Lama. He said the Venusians landed on Earth with a cargo of humanoid creatures resembling serpents and crocodiles and having human hands and feet. One of these reptilian creatures in fact, tempted Eve, perpetrating the fall of mankind. I remember, too, that this was also the time, just after my third trip to Iceland, when I received a confidential letter from Ritter von X and shared it with a special friend. It was the same nagging disturbance I recalled when I first watched the Dr. Stranges's video about a prophetic and benevolent alien from Venus, Mr. Valiant Thor. He warned us all about the One World Order and the Invisible Government—and reminded us of Valiant Thor's warning to Robert Kennedy before his assassination.

It was the same nagging disturbance. Just who was Valiant Thor, anyway? I wondered. That thought had disturbed me in 1976 and disturbed me much more in 1984. It still disturbs me now. I expressed my frustration on the phone with my neighbor, Wanda, who had become both my sounding board and my partner in research.

"This has got to be another one of those shaggy dog stories, don't you think, Wanda?" Speaking of dogs, I really need to get a grip on reality, I decided, as I continued telling Wanda my story.

My body beside Katrina's metamorphosed into that of my dog. I simply entered his form and attempted to chase the annoying female entities away. I growled loudly and ferociously, and even barked at them. I started to laugh because it seemed the best way to get rid of these annoying (but friendly) entities. Without realizing it, I woke up Katrina next to me. She said, "What are you barking about in your sleep?" I told her that ghosts were in the room and that I had changed into the form of our dog and chased them away.

All the while, I was laughing like crazy and observing the whole experience with detachment. It was an interesting experience—to say the least. I began to understand and respect the deeper meaning of Shamanism.

On the twenty-third of February, I dreamed that I had returned home very early one morning and was inspired to dig in a certain place in our front yard. I found many gold coins and bars of gold. While collecting the gold, I discovered a small wooden chest filled with a treasure of shimmering jewels. Eagerly I carried the gold and jewels into the house, excitedly telling Katrina about the magnificent treasure buried in our front yard.

While explaining my rich find to her, I heard a clear, distinct voice say, "*This is not the way of the Buddha to collect and horde gold.*" Katrina's cold and piercing eyes glared sternly at me. I dropped the gold in fright. Instantly, the value of the treasure changed for me. I removed a diamond jewel from the small chest, looked through it, and saw my future life in a vision of beautiful tropical trees. It reminded me of my mother's jeweled cross, which held in its center a jewel with the inscription of the Lord's Prayer. I peered deep into and through the jewel's timeless eternity and saw each day of my life as a string of pearls strung together, forming a mysterious path into a greater light. Then, in the jewel, I saw an illuminated core that was timeless. I saw an ancient language that only my subconscious knew and could read.

The morning after this dream, clear images of golden light began to form in front of me and spiral around my body. Whatever else I saw, this serene splendor radiated around me, and its golden peace engulfed me. This golden light-energy remained with me throughout the day, increasing its strength and power by small increments. My thoughts centered on this profound peace. My words gave me a sensation of power. Later that evening, Katrina and I were visiting my parents. Still everything remained within this radiant golden sphere. A soft, feminine voice entered my mind with these words, "*Please don't leave me.*" It was a spiritual voice that clearly wanted me to remain in close contact and not to leave her without guidance and direction.

The next morning, an increased surge of energy centered upon my forehead. It was so strong and intense that I radiated a cosmic energy that no one could deny in my presence. This energy was the center of my whole being; it gathered and moved all things into a supreme harmony. This infinite power was such that every event of the day was directly influenced by it. All my thoughts were of love and prayer. I was united because of the presence of this great light. I directed my attention toward the source of this golden light. While driving, I looked at myself in the rear-view mirror and saw a golden hue surrounding my head and face. In a solemn voice, I said, *"Within this light, I see no death, for its veil has lifted."*

I perceived many negative forces around me, which began to hover close. I knew these forces sought transmutation, to be released from their own negative condition and form. Somehow, I believed it to be my job. With this divine principle in mind, I realized that these negative forces must be confronted and met with the law of love. *The Law of Love* by Eugene Ferguson says: *"No matter under what circumstances, always meet everything and everybody with love."*

Wanda broke in at this point. "How many books do you suppose you've read over the past sixteen years?"

"Wanda!"

"I'm serious, Danny."

"Probably—I don't know, maybe—hundreds!"

"You should have a PhD by now."

"Okay, well—let me get back to the rest of it. A harrowing experience took place at Roaring Camp Park in Felton. This was where my blissful peace ended."

"You call what you've been going through 'blissful peace?'"

"I get your point," I chuckled. "But it really is important."

I knew this profound peace could release anything negative. I knew that all things around me were dependent upon that serene life for their very existence.

Katrina was angry with me. She did not understand that this cosmic peace was subduing and beginning to neutralize the negative forces around me. She, too, was seeking transmutation from the forces of a negative prison, just as I was seeking release from the negative conditions within me.

"Poor Katrina." Wanda sounded as if she meant it.

"Right there in Roaring Camp Park," I said, "I began to speak loudly to Katrina while people nearby were staring and listening. I attempted to explain to her that this golden light was transmuting all of these negative qualities from our lives together. I promised her that we would live much more happily because of it; that we would live in a much different, improved understanding of all that we would encounter in life. During this entire experience, I had the sensation that I was wearing moccasins."

"Moccasins?"

"Moccasins," I told her, matter-of-factly. The warmth of the Earth enlightened me through my feet. The Earth was as sensitive as I was and I knew that it, too, was a living being. A soothing memory of an Indian surfaced in my mind. In death, there was only darkness but in the light, I was reborn again and again. A radiant power moved up through my feet."

As I spoke to Wanda, I remembered another time—in 1982, just after my first trip to Iceland. A friend had given me the Hopi Indian book, which pointed out parallel legends regarding the center of the Earth. *From the Beginning of Life to the Day of Purification*, it was called, and it spoke of acts of prophetic consequence. The group leader, a chief of great wisdom from the Bow Clan, had disappeared into the dark night in search of the Earth's center, where clever, ingenious people from all nations meet to plan the future. The Bow Clan chief had found the place and was welcomed by beautiful people to an even more beautiful place where good food was laid before him. When he returned, his people understood that, by this action, he had caused a change in the pattern of life cycles of this world. Those gifted with the knowledge of the sacred instructions would live cautiously. The fate of the world would rest on their shoulders.

As I told Wanda about the radiant power of the Earth moving up through my feet, it connected—it all became clear.

My experience allowed me to become conscious of the Earth as a living, godlike entity. All the negative surrounding darkness was about to be released by the magnificent power of this golden light. I had a vivid image of a young Indian man. Through the memory of his experience, he explained to me how he used peace and harmony to transmute all negative conditions in our past and present lives. He described how the fall of consciousness repeats itself in rhythmic cycles until all negative conditions are eliminated by the presence of this golden light.

At that moment, the Earth was a living being that warmed me

beneath my feet with the power of all its forces. I knelt down, touched the Earth with my hands, and really felt it for the first time. The experience was like a baby first touching its mother. I was touched in return. The Earth and I were one and the same in the golden light. I fell to the ground and began to cry out loud to God my Father, who resides in his temple of golden light. I began to hear the Deity that created the heavens above and the Earth below and the Earth within. There was no other place to go. I thought of walking across the desolate stretch of Earth to some familiar and destined location in Montana. This Indian memory became a part of me and seemed to know where he/I had lived before. I was in search of my God. I wanted to unfold His will and make His presence known.

I had a wild urge to strip off my clothes. I wanted to remove the garments of time that enveloped me with horrible memories of my present and other former lifetimes. Everything began to swirl inside of me. People were intently listening and watching my strange performance. Like the old Indian bending over on his horse in that movie, *The End of the Trail*, there was no place to go, no haven of refuge. My trail of tears had come to its end.

I knelt upon the ground. The Earth turned into a shadowy, cold grave. I cried for a few moments while I slipped away from Heaven's golden light above. It was like a beautiful day of sunshine that had suddenly changed into a horrifying night of terror. I was entering a cave of darkness filled with all the negativity of the world—death, disease, and evil. Dark, cave-dwelling entities attacked me with their terrifying appearances. It was the world of the dead, a tomb of total darkness. The Earth and I were like an empty tomb filled with the memory of death, war, sorrow, and despair. It was the world of lost consciousness. It was the world in which we humans live.

This is what I believe the Greek philosopher, Plato, meant concerning his "Allegory of the Cave." In other words, Plato is the philosophical voice that tempts the man of the cave to join him in a journey to enlightenment. He views most of the population as prisoners, chained in a dark cave, watching shadows dance on the wall. They do not care what causes the shadows, nor do they try to resist the chains that hold their head in place. If someone were to break free, they would see that the shadows are the effect of people dancing around a fire. If one dared to venture even further then they would eventually find their way out of the cave and see the sun and the ultimate truth—for the first time.

I felt myself falling into tremendous disarray. I felt as though I had been viciously clawed to death by angry cats. I walked over to the pond outside the gates of Roaring Camp Park.

I stripped mental layers and old images away from myself and entered into my conscious lower self—wiser because of the experience of the precious golden light. I sat down near the pond and watched the ducks move swiftly across the water while several ducklings followed in their wake.

A young man about seventeen approached me and began talking to me about the ducks. He sat next to me and spoke comforting things about the invisible force that drove the baby ducklings to clamor for food from their mother. Then they grow up, he said, and become free from all attachment to their mother. He made me realize that I must let go of my baby daughter, Zephera, so that she, too, might walk the path of love that would lead her to a greater light and greater wisdom. I touched upon a world where all things in nature were in harmony. I had returned to our familiar chaotic world of the senses but with the memory of a cosmic treasure. This treasure is the Christ consciousness that touched and enlightened my reason, emotion, and intellect. This beautiful light encompassed all my thoughts. The stronger I grew in its peace and power, the more the darkness, in all its malice, would try to attach itself to me. I fervently sought the light to transcend my own limitations.

Late that summer, while Katrina and I were on a weekend vacation in Mt. Shasta City, a dream took me back home to Felton. In my dream, I was driving on a road that followed the contour of the river from Santa Cruz to Felton. As I drove, I observed true peace-loving people who were being nailed, crucifixion style, to the huge sycamore trees. Other people were beating them with whips and torturing them cruelly.

Their cries and moans pierced my ears. The tormentors who were whipping people ran after my car with savage, inexplicable anger. They wanted to crucify me, too, but I fought to awaken. I managed to escape the assaulting mobs, but not my fears. I woke Katrina from her sleep.

"A vicious beast from out of the bowels of the Earth has awakened," I said. "It is out to slaughter and to enslave all the freedom-loving people of the Earth."

I paused. Wanda waited in silence on the other end of the phone line. I suddenly remembered the Red Lama's *Aghartha*, which so eloquently presents the caveat against all religion, eternally used by the very evil agents from which it claims to offer salvation. The reptilian Venusians have domesticated mankind in our earthly corral in order to devour them at the appointed hour. This will happen when the gates of hell are opened by the very humans who were deceived into believing that evil could also be good. There is no safety in amassing war machinery and "peace in our time" is fast becoming a cliche'd illusion.

"What are you thinking about, Danny?" Wanda whispered. She sounded fearful.

"Oh, nothing. Never mind," I continued, without explaining what had just flashed through my mind.

I was alone in the dawn of a waking consciousness, and I needed to find my way back again into the greater light. It was my responsibility to alter these catastrophes and conditions upon the Earth.

I recalled and then relived a time in March of 1976 when a great force returned to me. While asleep, I turned to the inside of my body and put my hands together into a prayer position. A powerful force of energy surrounded my body. I was a mediator between my outer self and inner self. I uttered a prayer and summoned the name of God in Christ's light.

Someone spoke through me and moved my lips. This entity was completely conscious of my mind, body, and soul. I spoke loudly and with authority. I said, *"I am the All within the pyramid of you."* The voice was so loud that I awoke and again frightened Katrina.

It became increasingly clear and obvious to me what my life's mission here on Earth was to be. Those who had come into the Light were here to execute and participate in the cosmic plan.

Just as I was later to learn in the Hopi Indian story from the Bow Clan chief, it begins by raising the individual spirit out of bleak darkness so that a people and their planet may be ultimately delivered.

Chapter Fifteen

～ The Quest for Initiation ～

Oh God, who art the truth, make me one with thee in everlasting love.
—Thomas à Kempis

During that first week of March, 1976, we had guests whom Katrina introduced as some people from a nearby Christian church. It was her attempt to right all our wrongs; she wanted us to study with the church to help reconcile our terrible marital problems. I immediately agreed. The church wanted me to sign a contract agreeing to save our marriage and to overcome the forces of evil. The contract was written on an orange form inside a teaching booklet entitled, I Am Somebody.

I was asked to state my personal quest; I chose to know God's will for us. I wanted to be as open as possible. The contract included a place to add another person's name in order to help that particular individual. I signed my own name. I wanted to help myself. God helps those who help themselves, I thought. The contract also required another signature from a church member to witness our personal quest. Needless to say, I was ridiculed for my outrageous request. During these initial steps, I did not connect very well to Katrina. My quest failed with the church member. I just raged on and insisted that God helps those who help themselves. There was nothing more I could do to save Katrina. She needed to save herself; I told her that. Only the will of God was important to me.

While this was taking place, more uninvited guests knocked at our door. I invited them inside, knowing all too well that their mission was to convert me to their faith. Without hesitation, they entered and quickly occupied the living room couch. I took a moment and introduced them to the two young Mormon gentlemen, who were comfortably seated as well. Neither one really wanted to see or convert the others, but all of them most definitely had a vested interest in me.

I broke the silence, saying, "Each religious vision believes in a power that links their particular vision to reality." I told these witnesses of God that I was a lot like Joseph Smith, the Mormon prophet.

They mumbled in an awkward manner to each other and asked me, "What do you mean?" I faced my uninvited guests and said, "Joseph Smith flipped through the biblical pages, found a particular verse that suited his personal quest, and prayed earnestly for wisdom. He wanted to know which church to join. He received a vision of two angels and was told not join any of the churches, but to continue praying for wisdom. This is where you find me—between the differences of all religions. It is by knowing and understanding these differences that the light of wisdom will dawn as enlightenment. This is called cosmic enlightenment. It is because religions seek a truth, and many say to everyone, "I have found it and now you must know and become what I know. You must follow no other." Then I said, "The big difference between all of us is that some of us deny the Christ as the Son of God and believe He was never resurrected and that He was just a wise prophet. Others say, 'I'm seeking wisdom's light to find the truth. My choice is to have complete fellowship with God. That is the wisdom I seek.'"

None of them wanted to leave first, but I felt they wished they had never set foot into my house. I was simply stating in a humble way that I was petitioning my inner self, the god-self within. I paused and continued, "And how do you claim to be witnesses of Jehovah? That's a bold claim, to say the least. Am I being blasphemous if I tell you that I have witnessed Jehovah's light? Religions are great warring factors in my mind. But if you would seek the wisdom to know the differences, the human race would find the beginning of unity. If you unite those two polarities, war would end in the world and peace would dawn. Churches are congregations of all sorts of people from around the world. They sought those who were enlightened by God. Could these people be you? You asked to be enlightened by prayer and found the church that best suits you and your family. I'm not seeking a religion. I choose to find God's light. Religions do not bridge conflicts between themselves except through world war and world terrorism and ignorance and just plain hatred. If you have true fellowship with the Creator, why would you need a religion? Another great difference here in my living room is for you to interpret the Biblical meaning of Revelation 13 as the system of things. The well-known and accepted Christian believes that a certain individual will arrive and come to rule the Earth in a ruthless manner. Can't you see how I'm drawn into this diversity?"

Then came to my mind's eye the small, brown, snakeskin-bound volume, *The Science of Being* and was highlighted with gold letters and swastikas that marked the absolute laws of the universe. It was written with intricate wood-block illustrations, in 1923, by Eugene Ferguson. I remembered the blessing I had received on Mt. Shasta. A woman came to my small motel room, knocked on the door, and gave me this book. She said it belonged to me, and it was about the laws of the universe. It had been given to her by a very old man who wanted me to have it. He said it would prepare me for my ascension. This was what I had been looking for. She asked if she might exchange it for something of equal value. I gave her a book about the preparation of the ascension dictated by St. Germaine. It contained an unconditional form of truth about the laws of the absolute. All the physical and spiritual laws for the enlightenment of humanity were written there—answering all questions. What would it have told me in this situation? The law of polarity and duality is relative and transmutable, it said, above which God's love has risen. And we, too, can rise above the earthly law of duality. We, too, can rise above Evil, which is only the reversed upward movement of Good, it said. The various religious factions did not agree.

Religions simply told me—ordered me—to worship certain Gods, those that supposedly have the true power of baptism. Another example: the Catholic Church had emphatically stated that I was still married to my first wife and that I was presently living in sin with my second wife. I have always tried to humble myself between these challenges. I was in pursuit of only one thing: the light of wisdom.

I continued to question and asked, "Who should be given the supreme power in order to fulfill this universal religious vision?" They did not answer.

I questioned them further, "If I joined either of your churches, would this assure me the light of wisdom and resolve the conflicts between each church?" Again, they did not answer.

"I exist between your differences like a nebulous cloud. This is where I believe the light of wisdom resides and who do you say is worthy to behold that force?" Still, they did not answer.

Trying to be lighthearted, I told them that I want to join the church that God belonged to. I continued, "It is clearly known what our forefathers meant in that great document of freedom and rights for all mankind: 'Congress shall make no law respecting an establishment of religion,' reads the first amendment to the US Constitution. This clause has been interpreted to mean that the government of the United States—unlike Great Britain and other European countries—may not declare one religion as the national religion nor support one religion over another.

"The Constitution of the United States declares the right of separation between religion and state. Yet there is another great separation between church and God and, even worse, between man and God. There are those who proclaim God to be great while throats being slashed and heads severed. Yet those exist who toss out God, the commandments, and the Constitution, while martyrs reside in our country. In other words, convert—or die."

I suggested the idea of inviting a representative from each church to my house for a dinner and discussion. I wanted to have them all together so that each individual could briefly voice his or her origins, purposes, and goals—if they could. It could be its own kind of hell for them, but for me, it would further my search for wisdom. I believe wisdom has the power to judge all things rightly. This particular situation deserved a swift judgment that only wisdom's light could reveal. My purpose was to know the differences. I hoped to unite these opposites by prompting the light of wisdom and revealing a greater truth, but what truth?

I had shared this idea with my philosophy professor, but he warned me. He said it might prove to be foolish to arrange such a meeting since all of them would eventually make me their scapegoat and have me suffer for their narrow-mindedness. Needless to say, I never had my dinner party, even though the Mormons agreed with the idea. But the challenge was to be met in another way at another time.

One thing that all churches of the western world do have in common: they all believe in the scripture of the Holy Bible and that it will unfold in the way prophecy was written. Yet each religion has its own understanding of where humanity stands according to that prophecy. My concern is the fulfillment of that prophecy that remains undone. I found another agreement within the churches concerning the biblical prophecy of the Beast of Revelation. All the churches (except for the Jehovah's Witnesses) believe that the "Beast" is the world's system itself, not any individual.

With all of this whirling in mind, I went to fulfill a promise to my mother to tile a floor of their house. A great and wondrous light began to encompass me. My consciousness was rapidly rising and moving into a multi-dimensional form, and I was engulfed by a great cosmic energy. This energy began to envelop my entire consciousness while simultaneously expanding into the world's consciousness. All my attention began to grow in this light.

My now universal thoughts entered and encompassed the conscious world of Japan, then China, Iceland, Denmark, and Norway until the spiral of cosmic energy was whirling throughout the world. My mind's eye was opening and expanding. Remote viewing was at its best with detailed accounts. I could see by the grace of my mind's eye a clear and colorful picture of the cities of those countries. My mental vision scanned the world's consciousness into one single sight. I ascended and expanded higher and higher, like a seismic wave increasing in exponential magnitude. My consciousness rose higher until I was connected with all humankind. I bonded with this universal consciousness as one conscious vibratory force.

I was passing upward like a giant wave into a very high and audible frequency. The more the magnitude of consciousness grew, the higher I rose until I peaked into a pyramid-shaped awareness that encompassed all things of the Earth. At the peak of this consciousness, I could view the lower degree of world consciousness from its apex to its base as well as the circumference of our planetary consciousness. The light began to instruct me and reveal the soul of humanity as one; it suggested that no separation between God and man exists. This light was a sublime grace and a peace that yielded a great cosmic strength.

A strange twist of events followed, phenomena that are nearly impossible for me to explain. Hopefully the reader will bear with my inept attempt to explain a series of multiple cosmic experiences that clash in my everyday life. I will try my best to describe how two worlds collided into one another. It was during these experiences, both the cosmic and the mundane world (the world in which we live) inside me crashed into each other. The following drama and experiences truly surpass my own intellect and explanatory powers. I remain unworthy of clarifying the experience, unable to fully share its meaning. Perhaps some things just need to remain ineffable, even though it may convince some that I am truly going off the deep end. Nevertheless, I will attempt a description to the best of my ability.

My seven-year-old son, Art, was craw fishing behind my father's house. He had accidentally snagged my Dad's friendly malamute in the nose with a large, three-pronged hook from which hung a bundle of aromatic bacon strips as bait. The dog snapped and yelped in pain. Immediately a rapid shift of my consciousness began to occur. My cosmic deliberation began to take form. I drove the dog to the emergency veterinarian clinic, where the doctor removed the hook. Again my consciousness began to shift as I saw my Dad become calm—as soon as his dog behaved as if nothing had happened. Then, as we approached my Dad's house, more shifts in consciousness began to occur. From my inner cosmic sight, people appeared as past incarnated memory forms.

I viewed the world consciousness from top to bottom. It was like being in a movie theater, viewing myself. The light was the producer and I was the director. The inner spiral consciousness began to make its descent and I began to view my past, present, and future. It was my mind's eye that made the observation possible. I saw subconscious patterns in many people. However, not all people appeared this way. Each person was a unique and complex swirling geometric thought form. A short walk through the time and space of each individual furthered my knowledge of their past and future. My perception was keen. I directed these vibratory forces that were before me. The present was a window to each individual's past and future. The past, present, and future ran concurrently; there was no time—as I knew it. A division was inevitable because humans digress from their inner beings and begin thinking of themselves as separate, unique individuals.

Just then, I received a phone call from Katrina. She was going to the movies with Sharon, our next-door neighbor. I was skeptical about the invitation. I went home right away. When I arrived, Katrina was preparing to leave and wanted me to look after Zephera. I was puzzled. My impressions told me that she really did not know where she was going and that I was being deceived.

This had all happened before. It was a déjà vu, except that I was observing myself as one of those subconscious patterns participating in an event of mass cosmic changes. I was observing Katrina as a person caught between the present and her own past karma. I, too, was a part of the past karmic pattern. I entered into my own tangled web where the art of deception flourished. My mind held a dichotomy: How was I to maintain my earthly awareness at the same time as I observed my own karmic past? It would be by the grace of my higher self. I saw myself from within my own mind's eye, through a dark, hollow spiral, with only dual pinpoints of light visible in the far distance. It was like being trapped inside a great serpent, looking out through its eyes. The spiral was moving rapidly. I was drawn into a tight center within a collapsing circle. My consciousness was descending almost too rapidly for recollection. I was exposed to keen images of our past incarnate memories.

I felt a thick web of lies and deception from Katrina. Nervous and perplexed, I sat down to eat my cold dinner, silently questioning the doubt in my mind. Katrina quickly stepped outside the door to our neighbor's house. I wanted to ask her some questions, but she was avoiding me. While gobbling down my dinner, I was compelled to follow her—I had taken the bait. At that moment, I entered my own personal dark karmic pattern and now had to play out its role and direction. I was about to face those changes that were to take place in my life. I approached my neighbor's house, feeling like I was entering a nightmare of living hell.

I knocked on the door and Sharon opened it. She was a beautiful blonde, a newly-divorced woman who always wanted more from me than I was willing to give. I had once asked her to do some typing for me, and she approached me aggressively. The following weekend was to have been the end of her six-month divorce proceedings, and she had not made love for at least that long. She wanted me to come over that weekend and make love to her. I asked her why she had not made love during the last six months before her divorce was final, to which she replied that doing so would be an act of adultery. My angry thought was, "Jesus, you stupid, pathetic, poor excuse for a Christian." I began telling Katrina about her, saying to stay away. But unbeknownst to me, they had become good friends. I dropped the subject.

Now, there at Sharon's door, I asked to see Katrina. I invited myself in and sat quietly on the edge of the couch. Sharon left, with blatant disgust on her face. Katrina approached me, with large, wide bracelets draping her arms and rings on almost all her fingers. Thick makeup covered her face, and her eyes were black with false eyelashes and heavy eyeliner. Another past memory fell before me and between our souls. We were caught in a past karmic wind that was blowing us toward an uncharted course—like being adrift on a lost, cold dark sea. A memory of an ancient destruction began to raise—even worse, destruction of our human souls. I was crushed beneath my subconscious memory; in fact, I was drowning in it. My memory descended into the grosser cycles that lay deep within human illusion. My consciousness was landlocked. I was in a living hell.

These images needed change and relief from persecution. To experience the light is also to know and experience the darkness. It is light that exposes and uncovers the darkness. To see the light is to free the human consciousness from the bleak darkness of the soul. To be enlightened is the first step. Overcoming the darkness is where the work begins.

"What movie are you going to see?" I asked Katrina.

Sharon, too, was eager, jumping at me with anger. "She's free, independent, and over twenty-one, and she has the right to do whatever she damned well pleases, if it means going to the show or whatever the hell she chooses to do."

"Anyway, I've already told you," Katrina responded in a bitter, sarcastic tone.

"What in the hell are you two talking about?" I asked, resenting their angry attack, and feeling demoralized. They glared at me mutely, so I left the house, slamming the door behind me and leaving the situation unfinished and unchanged.

"Where in hell is the truth?" I thought as I stormed back to my house. We were dealing with shadows—shadows from a remote past, shadows Katrina and Sharon were completely unaware of. They were too busy living the same situation over and over again. I was in a dangerous karmic situation. I was determined to expose the hidden lies and deception that ultimately enslave us. How could I separate my actions from my thoughts? How could I explain these actions of deceit that were digging a deeper grave in hell? Our personal hell was a torturous bondage of our own karmic straps of timeless darkness.

Arriving back home where Art was minding Zephera, I told him to grab his sleeping bag and get into the car. We were going to my parents' house for the night. Once we were all in the car and on our way, I realized I had forgotten our dog. I didn't want to drive the car back up the driveway for fear of being heard, so I sent Art back to the house to get him, but warned that if his mother saw him she would not let him go.

Several minutes passed and Art had not returned. I drove back to the house. I walked in and Katrina was there, speaking to my mother on the phone. My mother persuaded all of us to come over and talk about the situation. So Art, Zephera, Katrina, and my dog got into the car. For the moment, everything was calm.

About a mile down the road, I noticed the gas gauge was registering near empty. Neither Katrina nor I had any money for gas, so I turned the car around and drove home. When we arrived there, we saw Sharon coming out of our house. Puzzled and angry, I yelled her to keep her Christian hypocrisy out of my home. Katrina threatened to call the police because I was yelling at Sharon. She made an angry attempt to grab Zephera from my arms. After several attempts, she violently pulled Zephera away from me, cutting her lip with the large copper and silver bracelets on her arms and wrists.

She went back to Sharon's to call the police, telling them I was acting like a madman. Another neighbor, witnessing everything from his front porch, asked me in a friendly, compassionate voice if I needed any help.

I nodded gratefully and we went back to my house for a discussion.

The police drove up my driveway and parked behind me. One officer spoke to Katrina and the other spoke to me. I told him that she had taken several valiums along with other painkillers for the dentistry work she had done. In fact, I reminded him that I had to hold onto her to steady her while walking into the house. Art had become frightened, witnessing her wobbly entrance. In addition, she was regularly taking antibiotics from her family doctor. She had been taking these for several months. I gave him an explanation of the circumstances that had led up to the present situation. He advised me to stay home and calm myself. Meanwhile, Katrina was to stay at the neighbor's that night.

In the next day's mail, on the 28th of March, 1976, I received a newsletter called the Pearls of Wisdom, No. XVII. It was dictated by the Ascended Master, St. Germaine, through Elizabeth Clare Prophet. She claimed to be an ascended master messenger from the Great White Brotherhood. To me this message was very appropriate at this particular time and moment, so I acted on this information and instructions, making a personal request in full faith to the Great White Brotherhood.

In the center of my front room, I prepared myself and sat in a mental triangle. This is the Shekinah, the quest for initiation. The Shekinah is the symbolic place representing the presence of God in our midst. It is the concentrated power of the Holy Assembly of the cosmic in the center of the temple. Once again, this request for initiation began to fill me with an uplifting consciousness. It was like a wave that raised my consciousness higher and higher. Through cosmic forces alone, I felt a perfect peace and balance amid an extremely chaotic relationship with Katrina and all our karmic ghosts.

I placed and unfolded the material in front of me and continued to read out loud. My mind was clear as I chose a personal path to traverse. On a blank sheet of paper I wrote down what I believed I am to know and all that I am to become. It was my future destiny. I was instructed to make two copies. I placed one copy between the pages of my Bible; the other was to be lit on fire and consigned to the laws of transmutation. This would send my message to the Holy Assembly of God, the ascended masters of which is a group of enlightened men and women who have conquered all earthly trials and reside in fellowship with the Creator.

This was done so that even now, at this very moment, this message remains as an echo secured in the everlasting presence of the most high. What I had written would remain a secret for the present; only in time would it be openly revealed.

My consciousness was overtaken as if by an endless ocean wave. It lifted me to an exalted power. This radiation encircled my mind, as though the power was the apex of freedom's light. This was the consciousness of what makes a man's soul free. This wave of consciousness rose higher and higher and I declared out loud with joyous excitement, "I am free!" I felt large in dimension and immensely powerful in my mind, word, and being.

I felt the spaciousness of the mountains, the sky, and the forces of the seas. All of these things were different waves of vibratory light energy. I could sense that all living things were nurtured by a radiant purple flame that also raised me into its omnipotent cosmic world. I placed my written request into a candle's flame. I watched while it was consumed by the peculiar law of transmutation and forever sealed. The only witness was the divine presence of the Holy Assembly. I was witness to the divine presence, and it witnessed me. One day, the seal of the transmuted message would be opened and it would be returned from the divine presence that had received it. I cannot speak of or predict the time or place or hour. But assuredly, as I am a witness to the greater light, the appointed time will come, and it will be witnessed by many. We will be received by the decree of Divine Law.

The following words flowed from my mind, and I wrote them down:

"Not as I am written am I known, but as I am written in your heart. I will speak through you and to the multitudes thereof. For in the beginning, your God spoke to Abraham, to be fruitful, multiply, and replenish the Earth, now you will know the Father of Abraham in them."

I was now faced with even greater opposition. My intuition was in full command, and I began to direct my thoughts. Man's selfish thoughts enslave others, but I was free and in a position to direct my actions, guided by the cosmic. This force was an expression of two cosmic extreme opposites. One force attempted to overcome a warlike force, while the other used compassion to achieve an equal power. These two opposing forces began to volley in my mind. They could be understood as supreme cosmic forces divided from themselves such as God and man, Christ and anti-Christ, light and darkness, or the first cause of creation and its devastating effect. The secret lies in the unity where man begins to have fellowship with his Creator.

For the next thirty-three days, I was attuned, by turns, to both man's world consciousness and the unpronounceable name of the Creator. It was an enormous and continuous struggle that demanded order and balance in all things. It was infinitesimal and galactic at the same time. The two opposites pulled mightily at each other. From the light of wisdom, I gained a glimpse of understanding into their destined unity. For now, however, these two great forces attempted to neutralize each other while remaining in great opposition in the mind of man. Both great forces were right; however, one was always trying to overcome the other. But their secret route to initiation was unity. This unity is the secret cosmic intelligence of all creation. It is what gave birth to the Christ consciousness.

It also is the first great cause of man. I was intuitively instructed for thirty-three consecutive days to sketch, focus, concentrate, and meditate on each letter of the Hebrew alphabet starting from aleph to tav. Each letter was a highly charged aspect of cosmic awareness. This was the only way to stabilize the unity of the first great cause of creation. I began to understand the DNA or blueprints of the tree of life.

I recorded a new calendar day in March. On the calendar I had drawn a beginning point of a new birth of consciousness. The flaming letter *aleph* formed and protruded through my forehead. It radiated throughout space in glittering yellow. I could see the letter flaming and dancing in front of me—then it faded. The sun had settled behind the mountains. The day grew dim, and then faded into darkness. Cool, crisp air rushed in through the open windows.

I stood in my son's bedroom and called his name through the window, across the yard to the neighbor's house. He did not hear me. I was not allowed to see him or Zephera. Katrina and Sharon had locked the door and did not want to see or hear from me. At home in the town of Freedom I was indeed free—freed by a flame of consciousness that endowed me with an authority to choose whomever I should serve.

Another day, a neighbor's son, Michael, came over to my house to show me his new Egyptian book about pyramid building. We shared the idea that the ancient pyramid was a structure for initiation. He added to the flame of consciousness and magnified its presence to yet a greater knowledge of the Creator. We discussed the Kabalistic meaning of God's unpronounceable name. It was about the Deity that once glanced upon the ancient Egyptian civilization, and that had reigned for thousands of years.

It was becoming very cool in the house. Michael wanted to build a fire in my fireplace, and so he ran outside to get some firewood. But he left the door open and never returned.

Darkness fell and a golden vaporous light began to glow and envelop my front room. The light grew intensely golden. It seemed to have come from the center of the fireplace. I sat in awe, viewing this wondrous light while facing the side of the fireplace. I called Sharon on the phone. "There's a beautiful golden light glowing in my front room, Sharon." My voice was as calm as the light.

"I don't believe you, Danny. Now, stop calling us."

The golden mist filled the living room with a soft glow. A new day was being created. Everything was in a state of cosmic peace. I wanted to find out where the light was emanating from. Something prevented me from going near or looking directly into the exact source. The light was spewing a radiant wreath of golden warmth. It quickened every fiber of my thought. I fell asleep in the quiet and cool air; I felt embraced by this cosmic luminosity.

The next morning, persistent Jehovah's Witnesses knocked on my door. They wanted to talk to me about God's coming kingdom and share the message about the world's system of things and the true understanding of the biblical meaning of the Beast of Revelation. We spoke about the meaning of God's promise to those who call upon His name. I felt myself falling into an endless abyss of religious powers.

The nameless God has creation under command, and we move and have our being within Him. At that moment, Ezekiel's vision came alive as I viewed the morning sky. It was a wondrous sight of four great forces blowing creation into creation. The four great winds sustained and moved the Earth on its eternal path. The Earth was not some junk heap but a vision held in the mind of the Creator. It was God's great cosmic plan. Truly a Paradise on Earth was coming to mankind in its entire splendor.

In my mind's eye, I walked into the providence of this great vision and saw a symbol of God's perfect victory. It was an enormous solid gold pyramid being constructed by humankind. It was a vision of humanity rebuilding the old world and replacing it with a new one. It was a new world without sickness and death that would endure forever as an earthly paradise, as it once was. This earthly paradise remains forever in the hearts of those who remember the experience of the light. Through the light the future is known. The world in which we live is dead and has become but a past memory to me. I live in between both worlds. Our present world is a haven of darkness with a speck of light leading mankind to Paradise Earth. The light is a reminder of who we are and where we are going. Looking back from the light, the Earth remains a dark pathway to the old world of pain, disease, war, and eternal death. This is the world in which we live.

The old world must be torn down in order for the new one to commence. That inner yearning at this very moment haunts humanity throughout the history of our world. The golden pyramid is a symbol and vision of God's plan. Not the god of this world, Lucifer, light bearer of the morning star Venus, but the God of all creation.

A wonderful thought of constructing a huge solid gold pyramid entered my mind's eye. I could envision it transforming into an eternal everlasting kingdom that would leave its mark on this planet for a time indefinite. It would be built upon the surface and in the hearts of those who would occupy its eternity.

The construction and purpose of the gold pyramid would bridge all humankind, and connect to the over-soul. It would lead us across and beyond the abyss of darkness—ignorance, disease, death, and into a most wondrous city of light right here on earth! Only light exists inside the interior of all matter. The pyramid would be a symbol of the cosmic mind with the golden light as the eye at its apex. The combined nucleus of all one hundred and forty-four elements of matter is the very center of the Earth and the eternal light of our beloved Creator. The nucleus is the center of all planets. The light is an expression of all matter within the cosmos. In the interior of the Earth our greater being dwells. The greater light is an invitation to leave the old world behind, as we know it. We must overcome the inertia of the earth's darkness.

I concluded the conversation with the Jehovah's Witnesses and spoke about many of my experiences that frightened Katrina. I told the lady Jehovah's Witness that a divorce was dead ahead for Katrina and I. She felt obliged to highlight a few Bible verses that would help reconcile our marriage and our discussion concerning the Catholic Church.

I was determined that Katrina and I should come to some spiritual understanding about our cosmic mission on Earth together. The next day I embarked on a campaign to save our marriage. I did not want to break the covenant I had made with God in the beginning of our marriage. I went to speak to a Catholic priest. However, about two months prior I had also spoken to a priest in Watsonville, who gave me a general understanding of the hierarchy and history of the Church.

I remember him saying, "If the Catholic Church is doing anything wrong, then we certainly have it coming to us."

I had asked him why Catholics were unable to become members of the Masonic Order; his reply was that the two are diametrically opposed to each other. I felt frustrated.

With that in mind, I returned to the church to speak again to the priest. A young priest there told me that the one I had seen earlier had been transferred to Monterey. I called him at the rectory at Monterey and I reminded him that I was the person who had asked, "What do you mean by two diametrically opposed orders?" He did not remember me.

Unhesitatingly, I asked him again what he meant when he said, "If we are doing anything wrong, we certainly have it coming to us." Again, he said he could not remember me.

I began to ask him questions that came from my own heart. I asked, "Why do people come into this church, call you Father, and kneel before the crucifixion of Jesus Christ? And why must we worship the God of the dead when our God is of the living?"

I challenged him on the phone. I said, "Why do you fool the people by being called Father and bow before an idol of a dead Christ?"

He said nothing; the next thing I heard was the phone being hung up.

I calmed myself and attempted to explain my concern about saving our marriage to the young priest. He did not seem to grasp the significance of my words about the cyclic repetition of divorce by the continual cause of adultery. This is the cycle that destroys the human soul. Breaking free from this cycle means the Second Death would be light upon the soul.

I left and went to find someone from the Masonic temple. I wanted to speak to anyone who could give me a Masonic opinion about why Masons are "diametrically opposed" to the Catholics. On the way to the Masonic temple, I walked through what appeared to me the shadows of the dead. People looked like the walking dead, their faces like hell fires were burning in their minds.

I noticed a flag at half-mast and asked an old man who it was that had passed away. He told me the mayor had died. I continued my walk through the memory of the dead in Christ consciousness. I pounded on the door of the Masonic Temple but there was no answer. On the bulletin board I saw a phone number for the custodian. I called that number and did not hesitate to share the questions about the Catholic Church, assuming that the custodian of the temple was a Mason. He told me to speak to his wife, and I shared the questions with her.

She said, "We are Catholics and we want to be left alone."

I asked if there was anyone else that I could talk to. She advised me to go to the Catholic Church and talk to the Monsignor. I told her that I had just been there but she only repeated her advice. I explained the whole thing again. Again, words were heard, but without understanding.

I immediately went back to the church. The question now in my mind was why are women and daughters throughout the world being used for fornication? I pounded on the door next to the giant church. On the way back, again I noticed the flag was all the way up, not at half-mast. Nothing seemed peculiar about death. Then, I found myself walking backward into time. Time is duration of consciousness.

"Thou art with thee," I heard repeatedly.

"Do this in remembrance of me," I heard as I entered the doorway.

After asking me politely to be seated, the priest remained silent. I began telling him about the impending divorce with my wife. I told him that I had failed my marriage. My quest to live a spiritual life had turned my marriage upside down. I explained my need for the clarity of truth that keeps a family and marriage holy and sacred. I told him that I was determined not to break the covenant I had made with God. I grew calm.

It was the grace of light that relaxed me from within. I asked the priest in a calm voice, "Why do people come into this church, call you Father and kneel before the crucified, the dead Christ?"

Before I had finished speaking, I noticed a soft glowing golden light on his face. I had uncovered and removed a veil, and the truth was exposed between us by this light. There was no darkness between us.

I stopped talking, and for the first time I understood what was meant when two entities are diametrically opposed. Because of our desire for freedom, wisdom, and love, truth unites all opposition and is witnessed by the presence as light. Our opposing knowledge was transformed into light by a greater light within us. This unity is the light of the world.

It is in this world of light that Christ made his ascension into and descended into the heart of the Earth. The truth was being spoken and the golden light was the witness and unity of that opposition. It is by this unity of love, life, and light that portrays a profound expression of divine grace and peace. I was a witness of that light. This was God's answer to us. To be a witness of God one must know His presence. A profound peace grew over my whole being. I had not been forsaken.

The priest and I stood up together and shook hands. "God bless you," he said.

I returned the blessing to him. The Christ light, I thought.

"Thou are with thee," I heard.

I went back home. I saw my son playing football and was reminded of the importance of freedom and how much one must sacrifice for this consciousness. A light of freedom encompassed my heart as Art called for me to play football with him. His soul carried this light of freedom. Again, a sudden rush of freedom's light entered my mind like a wave that soared high into the air. In that moment I had a sense of duty to defend that conscious freedom. A power was flowing through me, giving a deeper meaning to the understanding of the "I Am That I Am" consciousness. This is the conscious meaning of God's unspoken name. It is freedom's light that is the foundation for humanity to live and aspire to the inner dreams of each person's personal desire. This is a consciousness where the will of God can build upon and impart the plan of man and the destiny of his soul.

I began to feel this freedom awareness. It was like being conscious of all space. Continuing the quest for further enlightenment, I began to wade through this immense flame of consciousness. I walked through the test and trials of this immense flame of freedom consciousness as a pendulum that swings across an area of open space.

Katrina came over from Sharon's house and asked Art to return with her. I was happy to see her again. I could see a flame of freedom's light stem from her soul through her blue eyes. The flame was tinged with purple light. But she avoided me and ushered Art away, fearing something that was unknown to me. She was friendly and unkind at the same time and was not willing to carry on a conversation with me. It is an open door to Paradise to witness someone in the light but a living hell to encounter their darkness.

I began to feel very sad. My family was gone. I was one in God's light of freedom but I began to cry because the house was empty and my family was gone.

"Here I am, and here I am. I am free, and alone," I cried, asking, "My God, where is my family, in Your name? Look at me, Father, I am free in the midst of Your power and Your name, but where is my family?"

I calmed myself by taking a hot bath. I thought over recent events, knowing that I would never again be able to share these experiences with my wife, my family, or anyone.

I wanted to see Katrina at Sharon's house but the door remained locked. They did not want to see me. I phoned and talked to Sharon, who told me that Katrina had left and was not planning to return. I was sure she had walked to her mother's house. I was not alone during my search for Katrina, my daughter Zephera and my son Art. My search was one with the movement of all things around me. Everything was in accord with my search.

I arrived at my mother-in-law's house. She said that Katrina was afraid of me; moreover, she did not know where she was. I left and stopped a stranger, asking directions to the police department. The faces of the people reflected the golden light I had witnessed in my living room and upon the Catholic priest's face. I could see this light surrounding everyone—each person was being raised toward it. I could see it in everyone with whom I came in contact. The pathway was the light, and at each given moment it illuminated the direction to follow. The light was like a sword that cut its way through the darkness of my mind. The power of the light was magnificent. I was one with everyone in the spirit of this golden light. I was one soul in all people.

I proceeded to the police department, passing the Masonic Temple on the way. A crowd was standing in front of the temple. Arriving at the police station, I found myself traveling back into a familiar time. When I entered the building I was staring into a subconscious past. I told the two desk officers that I was looking for my wife and daughter, explaining the entire situation. I told them that my wife and daughter had left my neighbor's house and walked to my mother-in-law's. I emphasized to the officer again that she had never shown up at her mother's, and that I was extremely concerned. The officer called my mother-in-law's house and she repeated her story. I felt nearly psychotic—angry and hurt inside. I received the impression that no one cared and everyone was trying to avoid conflict with me.

In the heat of these discussions, three young and very beautiful girls of about eighteen years of age entered. Our conversation stopped in the middle. The officers showed more interest in the girls; they no longer had time for my situation. A peculiar scene unfolded as I watched an officer approach the girls.

One of the officer's attitudes actually began to form his appearance. I could not believe what I was seeing. Reality was changing right before my eyes. I noticed he was wearing a Hawaiian "aloha" shirt and large white beads around his neck. For the first time, I believed I was losing it. I went from present to future. I thought I was in Hawaii. The girls were laughing and manipulating the officers. If he was wearing beads when I had first spoken with him, I did not remember. Physically, he had changed into a different kind of person.

I felt a shift of consciousness take place. The whole picture was rotating from the past back into the present. The officer forgot he had ever been talking to me. I started to laugh at the situation and walked out the door. The girls followed after me, and I asked, "Who are you?"

"We are Job's daughters and we're just here to help," one of them replied. I thanked them for the distraction. They restored my sense of humor, and I drove to my parents' house and spent the night there. Of all the things I have experienced, that encounter was the strangest.

The next day I returned to my house to move the rest of my belongings. The thought of being a free man moving out of my house, abandoned by my wife, and having the children hidden from me did not make me a happy person. However, I was endowed with a certain comforting truth.

The following day, I went next door to Sharon's in an attempt to visit Katrina. She answered but was busy and still did not want to talk. I heard Zephera run to the door, crying, "Daddy, Daddy!" But Katrina refused to let me see her.

I returned to the house to finish packing my books and clothing and loading them into the car. I went next door again to see Zephera. Sharon answered but told me Katrina was gone and had left instructions not to allow me to see Zephera. Remaining friendly, I asked again to see Zephera, but her answer was still no. I understood the position Katrina had put her in, so I left the situation as it was. I was deeply hurt, knowing that Katrina and other people where holding Zephera from me.

I left my house for the last time. As a gesture of good will, I left a package of vegetable seeds on the bed of Katrina's eight-year-old daughter, Lesa. The idea was that she would begin to plant her own field of dreams. I left Art's football and his sleeping bag rolled up on his bed. He was left alone, unaware of the hard road ahead of him. After leaving these small symbolic items, I composed a note and left it on the kitchen table for the family to read and remember.

To Katrina, Lesa, Arthur, and Zephera:

> *In the times ahead may God deliver you safely to his presence so that we shall again be known in his Name.*

יהוה

I was living between two great opposing paths: the force and system of the anti-Christ and the unpronounceable name of God.

I ascended above freedom's light and felt these two great forces within me. The will of man opposed the will of God, and I remained suspended in a shaky, balanced state of awareness in my mind. I realized that freedom is only a preparatory ground where the cosmic acts. Freedom is awareness among the minds and hearts of humanity. Freedom is where God will establish His new covenant. The message simply meant to unite me with God's family and deliver that family into the presence of God's great kingdom. This was the message of the ascended master, Jesus Christ, to me.

It was made clear in my mind that the power of God would devour all governments of the Earth and set up a new kingdom—a kingdom that would not be brought to ruin or passed on to another people, but would stand for eternity. Mankind had only two paths to choose from—each was made clear. By the choice of free will, humanity would find its own fate.

My parents welcomed me back into their home, offering their unequivocal love and understanding. I felt empty and alone but was comforted by the familiar surroundings of their home.

Later that evening, my parents and I were watching a television talk show. On it was a psychic, a woman demonstrating a white light that was reflecting from the back of her head. Suddenly, I became aware of my father; he was a symbolic reflection of the most sincere, compassionate being of God. He reflected the glory of the white light with all its splendor of the Creator. His whole body was bathed in a sea of golden light.

I gazed at my mother. She seemed to represent an exalted glory within a vast, eternal sea of white light. The light grew until it filled the entire house. She turned about in front of me and a voice asked, "Please look and see me now?" Brilliant rays of white light from the eternal shone upon her, instantly restoring her youthfulness. She was the direct image of God in all His splendor of life eternal. With this glimpse of the eternal, her beauty was the mirror reflection and delight of the nameless God. Age was nonexistent, and disease and death were impervious in her presence. She was the reflection of the eternal with a vibrating energy that permeated her whole being. She had all the features of a glorious youth in the image of the Creator. She stood neither upon the present, the past, nor the future, but in the light of God's first image and reflection of his own being.

At that moment, a memory fell upon me like a shadow between times. My past relationship with my mother was during the time of ancient Rome, where I remembered a similar experience. For a moment I remembered the glory of Rome and its fall. Just as quickly, I was returned to the present, realizing that her journey beyond this life was to life eternal itself. Some people go to Heaven when they die, but others go to Paradise to live forever. This great fortune I saw was before her. She stood as the Holy Grail itself while I drank the gift of her living cup of life, held within the chalice of all time. Her journey of rebirth would end with a life in eternity for all time. She was destined to live in Paradise as Christ declared to the two thieves that were crucified next to him.

I then turned toward my father. He reflected the glory of the Creator. I saw in him as the root of all being. It was a wondrous sight to see his face reflected in the face of the Creator itself. We were one, but only one question arose in my mind. "Where is his face? Where is the face of God? He seemed to have secretly known that his own face veiled the Creator's. The past, present, and future were beneath my feet as a whirling disk of space and time, and the face of God was slowly turning to reveal itself again and set itself upon all humankind.

But that time was not yet to be. God's hidden face is a concealed plan for all mankind. A time was in motion to reveal God's full glory and His destiny for man. I could feel His consciousness change. The vision of His face came nearer and nearer to me with each moment that passed, but only God knows the time of His personal appearance with mankind. The universe was like an old watch that lay resting upon the palm of His hand. I was on a journey toward the center of all being, in a perfect and timed procedure that would eventually witness the face of the Creator. I felt joy—extreme and exciting. To me, this feeling and release of joy meant that all might know the Creator; that a watchful eye rests upon everyone and on all things to come.

Once again, space and time poured into my mind. I was a wheel within a wheel moving from the outer rim to the present time. Again a shadow moved between me and my father. At that moment I knew the fate and length of time of his present life. Intuitively, I knew that he would live in this lifetime no more than three years. That deeply saddened me; at the same time, his stature as he sat quietly portrayed a very ancient memory during the early stages of ancient Egypt. It was as though I had at one time known his face—a face that had once known the nameless one.

God's unspeakable name is the Lost Word that dwells deep in my memory. This is where the deeper mysteries of the universe are to be revealed once again. I continued my meditation as instructed by my higher self.

My father died of lung cancer two years later.

Chapter Sixteen

❧ The Second Death ☙

He that hath an ear let him hear what the Spirit saith to the churches; He that over comes, shall not be hurt by the second death.

—Book of Revelation, Chapter 2, v. 11

My quest for initiation continued. Later that evening, I removed the second sheet from between the pages of my Bible. I placed the sheet with a handwritten Hebrew name on it into the fire to be received by the law of transmutation. My mother placed another log into the fireplace while I remained transfixed by the rising flames.

For the remaining days, I was instructed to write each letter of the Hebrew alphabet on a blank white piece of paper. I wrote each letter, contemplated its meaning, meditated on its power, then set it on fire and tossed it into the flames. Every day, each letter showed me the power of its creative force, and every day had a story to relate. These projections of living flames were a unique aspect of cosmic consciousness. Each letter revealed to me a visible flaming color that projected through the center of my forehead. Each day revealed a new understanding—each day presented a significant new experience of cosmic awareness. Like so many of my experiences, these were so esoteric that words simply cannot describe them. To this day I remain in awe and without a logical explanation. However, I can say that it was a small glimpse into the DNA of the Creator's Tree of Life.

It was late afternoon. I was asleep on the sofa and my Dad was watching television. I was awakened by the voice of a talk show host on the periphery of my awareness. "Please, Henry, stand up for us, now." It was Henry Kissinger. I pulled myself up from the couch and stood at the same time as Kissinger stood up. I was one with his mind and knew his thoughts. He was inside my mind. For a moment he attempted and controlled my body movements and my thoughts. I could read and knew the force of his thoughts and mission, and it became my responsibility to alter his power and direction. It was my job to create a change in his mind. I attempted to steer his thoughts away from mine and wrestle his great influence around the world. Intuitively, I spoke through his mind to alter his secret plan. He was gaining too much power; it would tip the cosmic scale before his time. It became my conscious responsibility to neutralize his direction.

It was now May and the week of my birthday. I went to eat a late dinner at a favorite restaurant in Santa Cruz. I began to feel a vibrant light reverberating through my body; I parked my car and walked to the corner crosswalk. On my left, I noticed a white Cadillac approaching the same corner at high speed. I was so imbued with the vibratory light that I was unconcerned about the approaching car. As I walked closer to the street corner, the white light began to intensify and expand all around me. The speeding car and I reached the corner at the same time.

In that split-second I began to enter another place—not the restaurant I had intended. I heard a faint, inner voice warning me. I stepped in front of the moving vehicle without fear of death. I was about to walk into the mansion of my soul. At that instant, I turned my head and attention far to the right and up to the night sky. I caught a glimpse of a white light. I froze in that position and stared at a large, slow-moving white disk. It moved with grace and was large as a full-lit moon. Behind it trailed a long, elegant streak of light across the evening sky, from the north to the south until it disappeared into the night horizon. Transfixed by this unusual disk, I watched it continue until it passed over and beyond the redwood trees and the Santa Cruz Mountains. In a few seconds, the disk had glided softly out of sight, its white trail disappearing with it. My flight into another world was abruptly stopped and I was destined to stay and finish my job on earth.

My thought of the car was lost. I continued my walk across the street, unharmed. I entered the restaurant and ordered a vegetable crepe and a salad. It was nearly midnight. I finished eating, feeling I had been badly poisoned. My stomach became filled with an excruciating, spiraling pain. A frightening sensation filled my body, and I knew it had not been caused by food.

Without success, I tried to reject the idea that someone was making love to Katrina. I left the restaurant and staggered across the street toward my car. I felt short of breath, so I leaned on my car, resting my head on my clasped hands. I could feel her breathing inside my body. My face and body began to scorch and burn. An implosion was taking place in my soul—a chain reaction of multiple imploding stabs of fire. Then my body grew cold and dead inside, even while my soul was passing through a raging inner fire. It was the beginning of a living cremation of my soul. Thousands of stinging needles of death penetrated my face and neck with multiple sharp stabs of fire and coldness.

"My God, Katrina! Why are you creating death in my soul and body?" I cried aloud.

Somehow, I drove to her house. My heart and body were in a blaze of fire and ice. She opened the door at my knock and stood in silence as I walked past her into the bedroom.

I turned on the light. The man in her bed began to threaten me. He was naked, which prevented him from moving from underneath the covers. He was in a state of shock and embarrassment.

"Now you've really blown it," Katrina said in disgust.

I wanted to speak to both of them, but I was unsuccessful in my attempt to communicate. I stood there in silence, profoundly sad.

"Get out of here before you wake up the baby," Katrina yelled.

"I'm not saying anything, Katrina. How could I possibly wake up the baby? Zephera is better off sleeping than waking to this nightmare," I said on my way out the door.

At the doorway, I turned. "Why must my daughter live in this house with this going on?"

"Take it to court if you don't like it!" she snapped. "And stop coming around and spying on me."

With nothing more to say, I left the house, feeling rolling waves of white flames engulf my body, rise above my head, then overwhelm my body over and over.

A white fire consumed my face and head. My soul was being crucified and, at the same time—purified. I was the phoenix that rose from the ashes out of my own living hell. Then I felt numb from the purging, rolling white flames. Yet, simultaneously, a deep sense of compassion and forgiveness washed over me. It was a love that came from the very center of my being. In what seemed like forever in a single moment, I stood beneath the night sky and observed my nightmare of unimaginable pain. I was acutely sensitized to the spirit of the living Christ consciousness.

The compassionate feeling stayed with me, pouring through me like a dazzling light. A voice echoed inside of me. "Thou shall not commit adultery. To transgress this law, the Shekinah suffers the being who gives us light, life, and love." This is the meaning of the Christ self within man. The Shekinah is the symbolic place representing the presence of God in our midst. In other words, it is this light that sacrifices itself to erase the negative conditions brought on by the transgressor. By this reason of love, our inner being sacrifices itself willingly for our personal salvation. It has been made absolutely clear to me how to keep God's Holy temple of life eternally pure.

Without honesty, there can be no true love for our Creator. Without the love of God, we fall in bondage with no promise of freedom. My own heart was being purified. Love, freedom, truth, and honesty are just words, but the thoughts behind them resonate with the presence of God.

It is all so very clear and easy. Why could she not she see it? Honesty is the root that blossoms forth with love. The blossom of love assures freedom. Freedom gives flight to the soul. The soul unites our hearts as one and flies directly into the heart of God's wisdom.

I said aloud to God, "Please forgive all that we have done and hold us high in your favor, so we may yet have time to live your law, for I have seen where I have been." The fire moved away from my being. To my surprise, I became youthful, and this new feeling poured through my mind and heart. I realized that mankind is a living, holy temple and inside us is the fire of the living God, a God who wishes to bestow the gift of life eternal.

Through all my most sincere actions, I was supporting the highest morals that keep a family holy and sacred to God. I was defending marriage and its law because the law and its maker are inseparable. A marriage is the highest moral obligation and responsibility that confronts and tests men and women. I wanted to explain all of this to Katrina, but she would not listen.

Only the night sky heard my words. God was almost visible to me. His light empowered my heart, mind, and soul. I then spoke aloud to the God of my heart.

"I know we are always visible to you, but my love and compassion are invisible to Katrina. I am so overwhelmed with an inner love that I must support your highest morals and laws that govern a marriage and family. I know I need not fear judgment as long as I am planted firmly in your law."

I grew to understand that lawlessness in the eye of the Creator is certain self-judgment. Man's laws can be changed, overruled, and appealed but the laws of God? Never.

By living God's law, I found freedom. I felt God's freedom as a dazzling purple flame of consciousness that gave rise to victory in the name of truth. Freedom is God-made. God is free and master of all truth and in that truth I was free.

Freedom becomes an illusion when free will is exercised without having an inner understanding and relationship with God's law of love. I was quickly learning that freedom is a consciousness that becomes the foundation for God's will. This will is a force that operates within each individual. When a person claims the Christ light, freedom serves his mighty cause. The will of God is the law, and it is supreme over all. Without freedom, darkness will always prevail. Free will is a choice that either enslaves us or sets us apart from the limitations and bondages that bind us to the Earth. For many people, money and other material objects have been made into material gods that enslave them and keep them in bondage. Ignorance and fear support and feed this material god that will yet fall upon its followers and devour them. This material god that man worships has become a false image before its own kind. It is a false god. The god of this world holds people captive by disease and sickness, death and sorrow.

I asked, "What is behind marriage that has been kept holy and sacred by God's law?" Behind marriage lies the temple door to the Creator. Through the sacred knowledge of the triangle, marriage between a man and woman is a preparation for the ascension. Marriage is a sacred trust. Through this union between a male and female, they successfully rise into the knowledge and power of God and become agents to aid his creation in any capacity.

Marriage is an alchemical union—a process of purification by fire that prepares us for the final initiation into cosmic consciousness. This is the awareness and realization of God. This is the path where the cosmic tests us by trial.

This is the mysterious metaphysical path of the Christ consciousness. Once I had met and endured all the conditions of God's sacred and absolute laws of life, I was led to contact the Christ consciousness. This is accomplished in the life cycles of birth, death, resurrection, crucifixion, rebirth, and ascension. This is the great personal plan for each and every human soul. Once the soul returns to God, Paradise returns to man. This to me is the absolute meaning of Christianity.

I am a Golden Age man destined to return with many others to an earthly Paradise, surrounded by a glorious splendor within the presence of the living God. To me, this is the ultimate purpose of marriage: to fulfill the divine laws of creation. For me, anything less would only alter the truth. Without the love of God in marriage, truth cannot exist. Without truth, freedom becomes an earthly bondage. Our human destiny is to complete the universe and fulfill the divine laws of creation. We must fast and pray to bring an end to our old world. Fasting means to be baptized by water so that we can be baptized by fire, which, in turn, means we are ready to receive the glory of God's light.

"Will Katrina ever know these truths, God?" I pleaded, too weary to wait for the answer. I got into my car and I started the drive back to my parents' house.

Chapter Seventeen

❧ Breaking the Chains ❧

Who is free? The wise man who can command himself.
—Jonathan Swift

During the following night, I was abruptly awakened. Frightened, I found myself staring into the eyes of a large, ghostlike leopard. The wild creature roared fiercely as it approached me in an attack position. It leaped at me and entered my body, and began to claw and slash at my heart.

Knowing no other defense, I called upon the name of God. A bluish-purple light swirled into the room, entering through the top of my forehead like a laser beam. It pierced my heart and encircled the mysterious cat visitor until it was destroyed. Through my forehead I projected four familiar, dancing Hebrew letters that vibrated in a soft green light. I was calmed by this light and fell into a deep, comfortable sleep.

The midnight battles were not entirely over. My dreams for the following year were filled with images of confinement, deception, and destruction. I suffered through vain attempts to reconstruct and repair the damaged infrastructure of my consciousness.

Jimmy Carter had just won his victory for the presidency of the United States in November of 1976. In a dream, I was confined in chains. I saw Jimmy Carter, arms folded, with a cheerful smile of victory. I saw my brother David standing beside him in silent agreement, reinforcing his victory.

In this dream, David had a handgun and was pointing it at me. I was lying on the floor, my entire body wrapped in chains and my hands tied behind my back. My clothes were ripped to shreds as a result of the maltreatment I had received. I had been beaten and thrown into a small, locked room; I was a hostage. This was the vision I had under Carter's leadership. This dreamlike vision symbolized a warning of some kind—unforeseen chaotic events would befall the American people during his presidency.

The following year, in April of 1977, I dreamed my brother, Fred, and I were on a spaceship. Katrina was there only to witness my demise. I counted down the seconds as the numbers appeared to me on a clock—six, five, four, three, two, one—and then, crash! Thump. We braced ourselves as we came in contact with the moon. I opened a large circle-shaped door that exposed the moon's surface. We exited the spaceship and walked upon the surface of the moon. I submerged my hands into the soft soil. Water filled the holes that our hands made.

The three of us continued our exploration and came upon two stone structures. One was fenced with a sign warning us of an extreme danger zone; it forbade us to enter. Heeding its warning, we avoided it. The other stone structure was an ancient ruin.

We felt an urgency to rebuild it. Fred began exploring the ancient structure, while I was compelled to rebuild the ancient mound upon which it stood. It was similar to a pyramid but conical in shape. Someone else had built this structure, but time had nearly eroded the mounds into dust.

I walked toward Fred and saw four large creatures shaped like sunflowers bend and bow low to him. They had leaf-like arms extending out to him as if he were some kind of life-giving force. Surprised, I asked Fred if he saw what I was witnessing. His reply was an emphatic yes, but the tone of his voice suggested that I should not speak to anyone about what I had seen. We walked together and came upon four more sunflowers. These were tall and in full bloom, and had no resemblance to people. As we passed by them, they too bent over and bowed low to us. The huge sunflowers nearly touched us as we passed by.

Later that year in August, I was given another such visionary dream. It was a precious gift from the Akashic records—or my genetic memory, as Ritter von X would say.

My mother, David, and I were standing on the beach next to the Pacific Ocean. In the far distance, I noticed an island slowly rising from beneath the sea. We could see the various shades of blue water as it increased in depth, indicating we were in the tropics. The island was barren and covered with a thick, rolling fog. Four vicious, roaring tigers appeared out of the foggy mist. I could see part of a temple behind the ancient, long-toothed tigers. My mother and brother stepped back in fear of these creatures. They stepped into their own fear and into what seemed a lesser light. I turned toward the tigers without fear, saying they would not harm us. After I had spoken, the tigers halted their vicious advance and disappeared.

By the same light, I was able to walk upon the crystalline blue sea. I came to the front side of a temple wall and saw ancient writing upon it. I found myself near a tomb in the middle of a rock temple. The fog disappeared and I noticed that the tomb was open. I perceived that a living soul of great light had just left the tomb and had flown high into the warm, sun-drenched atmosphere.

While I observed this light-filled soul, a tall man approached me. He had striking blue eyes and was tanned by the sun. He wore a white linen loincloth around his waist. He stepped out from behind a shaft of violet-blue radiant light. This shaft of light reached far into the sky. The radiant man handed me a large golden bowl containing a bluish liquid light. He extended his hands toward me and offered me this cosmic liquid drink. Without hesitation, I grasped the bowl with both hands and drank half of the liquid. I spilled some down my chin and felt it drip through my beard and down my chest.

I stopped drinking for a moment, and then held the golden bowl toward the sky. Suddenly, I began to laugh hysterically. My laughter increased into a large echo and then released me from my fear and bondage. I turned and looked behind me and saw many other people drinking the same bluish liquid light from small golden cups. The people who drank from the cups partook in a most humble manner, as if they were in the presence of a great teacher. I was in the presence of a great soul—a giver of the light of freedom. This was a call for me to seek a new location, perhaps somewhere in the South Pacific, where the cosmic subconscious would rekindle my memory of the great lost Garden of Eden, and where mankind was first introduced to Paradise. This is where I witnessed my mother's return to Paradise.

This dream was a glimpse into my immediate future—a future that was certain to bring change, and an opportunity to break the chains that had kept me from the freedom I was seeking. It was a glimpse of my personal destiny into a mysterious enlightenment and my insight into the dawn of the Golden Age.

On September of 1988, more than ten years later, I recorded the following vision. At last, the chains seemed to have been broken.

I was awakened by a solemn silence just before the morning sunrise. It was around four a.m. and I had a serene feeling of peace and calmness. I sensed a presence. It was a perfect time for meditation and prayer. I pushed my pillows against the wall and sank my back into their softness. I pulled the warm covers up to my waist and closed my eyes. My thoughts were clear and I felt relaxed. In a childlike manner, I waited with a sense of excitement and anticipation. I began to recite the Lord's Prayer. The moment was God's. With silent words, I closed the prayer. My attention was now focused on the center of my forehead. I saw an inner darkness. Slowly, a funnel-shaped, wobbling tube began to form. I was drawn into its darkness. It began to swirl and form into a long, snaking tube. Next, I was traveling down its interior. I saw a golden pinpoint of light at the very end of this serpentine tunnel, and I began to focus on it. It looked as though I was seeing through the eyes of a hollow, spiraling serpent.

I noticed three-dimensional figures moving around on the outside of the spiraling tunnel. It was a reflection of my own making—an inner darkness and fear. I saw human figures moving about, trapped in the illusion of death.

I gave the darkness no attention. Instead, I kept myself focused on the ever-expanding golden light at the end of the swirling tunnel. Through this vast tube of darkness, the golden light kept growing and traveling up the spiraling tube. I remained passive while watching from one world into another.

The circumference of the gold light steadily grew in its force, brilliance, and power. The light was racing toward me. Without losing its intensity, it impacted and exploded on contact as it touched the center of my forehead. At that moment all darkness disappeared.

The next moment, I found myself standing outside my house facing the northern night sky. The starry universe veiled an immense world of radiant blue. This was the world behind the darkness of the known universe. At the northernmost point of this world of brilliant, radiantly blue atmosphere, I saw immense white clouds forming. The universe is a reflection of a greater and vast cosmic universe behind our own. These large clouds moved from north in an easterly direction in front of me. They raced across my view, across the open field, then up and over the tall redwood trees. These angelic cloud forms appeared to be about twenty feet away from me, and about twelve feet high. They came and went like heavenly angels, each pouring its radiance out to me with a divine touch of love. The cloud figures danced with a kind of joyful compassion, speaking a language in musical form. It is a vision of grace and beauty beyond any human emotion I have ever had. Each figure was filled with a divine musical presence that reached out to me with its entire loving splendor.

It was an invitation to enter another world—a world beyond human comprehension into a Paradise of eternal life. Almost instantly I was returned to the comfort of my bed, with this divine encounter forever etched in my memory. I slipped into a deep, solemn sleep.

Chapter Eighteen

∾ FIRE AND ICE ∾

Some say the world will end in fire,
Some say in ice.
From what I've tasted of desire
I hold with those who favor fire.
But if it had to perish twice,
I think I know enough of hate
to say that for destruction—ice is also great
and would suffice.

—Robert Frost

November of 1976 was a particularly cold month in Felton. In my dreams one night, I entered a mystical state of awareness into another time. I saw myself with eleven others in a wooden boat. Icy waves were tossing our hand-hewn, Viking boat in all directions. The boat was just offshore a newly-created island, known as Iceland, in the vast north. As our little boat approached this new island, only a dim sun lit the day, but the true sun was the power and light within our minds. The men on the boat were ethereal beings of light. They made the land rise from beneath the torrential seas.

The men moved in unison and harmony with each other. They were co-creators. The boat ran aground and they disembarked quickly. My feet touched upon the newly-created surface of Earth. Strangely, it felt warm—not icy, as I had anticipated. A multitude of volcanoes were erupting with rivers of lava flowing rapidly to the open sea. The lava did not touch us.

All twelve of us walked toward the interior of a mountainous volcano. We came upon a gigantic stone that blocked our path. We pointed our hands in unison toward this mountain-sized boulder as we made an intense vibrating sound. A command was given to move the stone and a familiar voice rang out in opposition to our request.

It responded with a distinct and clear, "No."

Nevertheless, our response was, "We must."

And, with that, the huge stone began to move. It rolled away from us and slid down the hillside, splashing into the sea. The island was covered with a thick, vaporous steam. Occasionally, the mountains were visible through the vapor.

Behind the giant stone was exposed a brilliant white fire that rose high into the newly-created atmosphere. At the top of the pyramid of flames were distinct letters dancing above in white fire. The flaming letters were of a white brilliance, intertwining within one another and seeming to separate from the apex of the pyramid of flames.

I recognized the letters. They formed part of the first language of the Creator and the vision of the Lost Word. When this language was spoken, creation occurred. Vibrant symphonic sounds, somehow familiar, were behind the dancing flames and its musical message was etched into my mind. The scene and sounds were the blueprints of Creation. It felt like a mythical time when, in the beginning, God was man and men were gods. It was the great law and the authority of the omnipresent being of creation.

I awoke the next morning with this memory still intact. Katrina was due to return that morning. She had agreed earlier to reconcile our marriage and once again drop the divorce proceedings. Thinking about the dream, I waited for her return, looking forward to a new plan and direction for our lives. Katrina did not return.

"She must have made other plans," I thought, and decided to call her.

"Please come and see me," she said. "I have to tell you something." Her tone was hesitant, skeptical.

I drove to her house. My entire being was overwhelmed with conflicting emotions, and my body began to vibrate with anger.

I knew she had changed her mind again and wanted out. The memory and meaning of my dream became clear and keen. God's law was etched in my mind. Within this deep, subconscious memory, I was uplifted and floating ten feet above myself. I felt as though a comforting friend—a big brother or a great teacher—was with me. It was like that being I had witnessed in my dream-state awareness.

Katrina met me outside her mother's house. "I'm going to be married," she said hesitantly, almost questioning.

My subconscious memory slammed into a forward motion, plunging me into the present. Calmly, resisting the anger that had been feeding me, I spoke to her of my dream. I did not hear my own words as I spoke. The people passing by looked like the living dead. She had entered a world of the living dead.

My mind was swept into a beginning of a new world. The old world consciousness was quickening and fading at the same time. I was invited into the house of creation—the mansion where God dwells. Suddenly I knew what was left undone in God's creation and began to understand the meaning of becoming a co-creator.

I understood what I was putting upon myself—it was to restore my consciousness to its original form. Katrina looked both beautiful and untouchable to me, while my desire and emotion began to flood—then shatter, flood—then shatter, in overlapping icy waves.

Once again, I was left abandoned and alone. She slipped back into an open sea of darkness. My memory was like a spinning wheel. I spun, ascended toward the heavens, and descended at the same time into the lowest pits of hell.

I asked and prayed aloud, "When will be this new Earth within man? What has happened to man that he has bowed to the order of the living dead? What has happened to the order of men who raise their heads high in the mighty presence of Your great law?"

I lacked the knowledge necessary to create a new Earth. I understood Katrina's attraction to the old world. The world lives there as well. I saw her fade deeper into the world of the living dead. I was the memory of the old world. She wanted me to understand her new life and to part as friends. I resisted her wishes and persisted in the attempt to make known to her what was being made known to me.

I was angry again. People passing by me became frightened.

"Katrina!" I shouted, "I want you to know about the vision and knowledge of the Creator and His mission of a new Heaven and a new Earth!" I felt as though I was traveling through Van Gogh's Starry Night, twisting, turning, and spinning into a raging cosmic storm of near-insanity.

Katrina did not understand or care to hear any more.

I chose to leave in anger, all my senses battling against her ridiculous idea of a new marriage.

I got into my car and left the vision of the old world burning in her presence. She was swept away into a world of destruction. I vowed to do all I could by God's given power to stop her treacherous direction.

I left her with these thoughts and questions ripping through my mind. What memory can man receive in order to make known the origin of the new Heaven and Earth? How much time do we really have? How does the Creator merge his thought among men and live? How would the Creator trust man with his knowledge? Is this not the reason for man's creation? Who am I that God is mindful of me and not others like Katrina?

My awareness flowed in and out like concentric circles, each containing a different memory in time.

I finally let go of the situation and decided to move from the country. I decided to buy a one-way ticket to Reykjavik, Iceland. I knew the answers to some of my questions would be found there. The reason was not apparent, but I was compelled to make a journey to the far north, where I might encounter a Nordic connection with man's destined Paradise. Why or how, I was not exactly sure.

I had always known Iceland would play some kind of important role in my life. To what extent, would eventually become known—I was sure of that. An unknown adventure of discovery awaited me. I was ripe for a change of environment, and set my departure date for December 2nd at eight in the morning.

About one week before my departure, Katrina called me late one evening and asked if I would visit her. Without hesitation, I left my parents' house to go and see her. Once again she had radically changed plans about her new marriage. She told me that perhaps our marriage could be reconciled after all. I agreed, but with some hesitancy.

I spent the night with her. The next day we began to look for a house. Once again our marriage was reconciled, and in my mind it was to be a family gathered into God's name.

I told my parents about my change in plans. Needless to say, they were not overjoyed with my decision—they were tired of all the waffling and confusion. They tried to interest me again in traveling, telling me about a television program they had seen about Iceland. They felt it would be in my best interest to venture out and find what I was so earnestly searching for. Their opinion was that I was better off without Katrina, and that I need not worry about Zephera. They felt strongly that Zephera would always be close to me. They preferred that I go directly to Iceland and satisfy my inner prompting. As my mother put it in plain words, I needed to turn over a new fig leaf.

That, however, would have to wait for some years, despite how we all felt at the time. In the meantime, my dreams and nightmares of societal and personal collapse would continue.

In February of 1977, the image of my old friend, President Pierce, visited me again in a dream. Again Fred was there. This time my brother was showing me some rare old gold coins. These coins were early American Stella three-dollar gold pieces. I examined the gold coins, after which he showed me an oval-shaped locket. I opened the locket, which contained a picture of President Pierce. The photo was preserved by a rare gold coin that rested upon it. I recognized his image and understood, remembering his spiritual quests. Fred told me that the coin was a counterfeit. The coin was only a symbol that represented the origin of our nation's system and order.

This meant that our political and economic system is the balancing point between the free world and the tyrants of many nations. Our country is preserved by an ancient memory while modern dictators strive to end the vision of Plato's Great Republic.

I was told to take these coins to a dealer and get them appraised—but not sell them. On my way to the coin shop, I met a man who was very excited about the rarity of the coin and the meaning of the hidden message. The coin dealer examined them without saying a word. A few moments passed by before he spoke.

"It is because of our partial knowledge and memory of our cosmic origin and our enlightened reason and intellect that our philosophy gave rise to the light of freedom and the system we created long ago. America and the free people of the world were created by following the natural laws of the Creator. When man remembers his origin, the destiny of mankind will materialize."

Then the dealer transformed and blended into a memory of an ancient time. He invited Fred and I and Mr. Pierce into the Akashic memory of Atlantis.

In awe, we witnessed the ancient architecture and giant temples. Pillars were carved into huge doorways—gateways into the slopes of gigantic mountains. The large island of Atlantis had been taken by fiery storm, leaving the ancient ruins careening into the sea to their final resting place. Mr. Pierce shared all that was recorded in his Akashic memory. I was grateful to have witnessed the deeper understanding about our human origin and destiny. Atlantis is the undergarment of our country, the United States of America.

Atlantis Imagined, by Lloyd K. Townsend

Critias speaks: Let me begin by observing first of all, that nine thousand was the sum of years which had elapsed since the war which was said to have taken place between those who dwelt outside the Pillars of

Heracles and all who dwelt within them; this war I am going to describe. Of the combatants on the one side, the city of Athens was reported to have been the leader and to have fought out the war; the combatants on the other side were commanded by the kings of Atlantis, which...was an island greater in extent than Libya and Asia, and when afterwards sunk by an earthquake, became an impassable barrier of mud to voyagers sailing from hence to any part of the ocean...Zeus, the god of gods, who rules according to law, and is able to see into such things, perceiving that an honorable race was in a woeful plight, and wanting to inflict punishment on them, that they might be chastened and improve, collected all the gods into their most holy habitation, which, being placed in the centre of the world, beholds all created things. And when he had called them together, he spoke as follows-

—From the Dialogue of Critias, written by Plato in 360 BC, translated by Benjamin Jowett

The rest of the Dialogue of Critias has been lost.

The powerful memory of this dream faded and I awoke knowing the secret destiny of America.

My thoughts were uplifted, knowing the United States of America and all free nations around the world remain the last great powers on Earth. The free people of all nations preserve the way to our destined Golden Age. Our country was brought forth by the hands of the enlightened ones through the ages of time. It was they who lit the torch of freedom and revealed the memory of the Golden Age. This glorious light of freedom preserves the minds that carry the Akashic memory of the once great and prolific Golden Age. The very light of freedom is the key that opens the door to the Golden Age. Are we not living the vision of Plato's Republic in the minds of all the free people? Was it not Plato who recorded the last days of Atlantis?

Freedom is a consciousness that protects and preserves our present and future way of life. Freedom is a path to a higher purpose and when this light of freedom has served and completed its divine purpose the need for any government on earth will cease to exist. People will live in complete fellowship with the Creator because divine law will be written within their hearts, minds and souls forever more. To lose our freedom is to lose our souls and all that we know in the light of our Creator. Just remember those millions upon millions who died so that we could be here today. Their deaths are our gift of freedom. Our oppressor is alive and well and wants to destroy the very likeness of freedom loving people. Remember! What truly guides and protects us and those who to this day give their lives so that we can remain free.

Chapter Nineteen

⚜ Future Lifetime ⚜

Conflict and war among nations begin with friction and disunity among individual people. A nation at war is merely the effect of a spiritual darkness born of intolerance among individuals who comprise a nation. When two brothers find reason to disrespect one another, or when two friends find ways to fault each other, then two nations will devise reasons for which to engage in bloody battle.

—*Rabbi and Kabbalist, Yehuda Berg*

EVERYTHING WAS FALLING apart and taking me under with it. I was buried again, like Atlantis lying in ruins at the bottom of the sea. Once again, I was living with Katrina and the children. After our brief separation we rented a house in my hometown of Felton. Two years passed. My brothers and I had a lumber import business, and we had been doing fairly well until interest rates hit record highs under the Carter administration. The business came to a halt. Egos, tempers, and plain dishonesty also sabotaged our business.

My father was dying of lung cancer. My marriage was taking another nosedive. Katrina was in the fifth month of a pregnancy she wanted to terminate. One evening she told me our baby was dying. She evaded my questions, saying she didn't know why—it may have been caused by umbilical cord strangulation. If I wanted to know anything else, she told me to go ask the doctor. I fell into tears with disbelief. Was it because of my exposure to the defoliant spray, Agent Orange, in Vietnam? Perhaps I will never know. Nonetheless, I felt guilty, as if it were somehow my entire fault.

That evening, I was struck by a quick jolt of gold light. Later while I was asleep, a beautiful, angelic being appeared and radiated another shaft of golden light upon me. It was a message from my angelic, unborn baby daughter. She was a beautiful female telling me goodbye and not to worry. Her message assured me everything would be all right with her voyage into the unknown. Her radiant blue eyes were more concerned about me than her own fate. Her name was Ayesha. The small fetus miscarried. Was it Ayesha herself who had made the choice? I will never know.

I watched the procedure. The doctor turned to me and said that I was right—it was a female and there was no apparent strangulation. Still, my heart was saddened. I felt deceived and lied to—again. The doctor never charged me for his services.

I was cast into a whirlwind of change. I had a dog, Ruffles, whom I had rescued from a car accident. The children loved him, but Katrina always chased him away. The dog would always run to my parents' house just under a mile away. Katrina refused to have him around the house and children.

"You have got to get rid of that filthy animal," she would say, chasing him with the broom out of the house. "Artie, don't let that filthy animal sleep with you," she would say. My dog Ruffles, a beautiful shepherd and coyote mix, just stared with unceasing love while forgetting his own pain. He had a brief hard life, having been mistaken for a coyote and shot while we lived in Montana. Still, Katrina seemed to take out all her wrath on this innocent animal.

One Tuesday evening, I attended the convocation at the Rosicrucian temple in San Jose. When I reached the mountaintop twenty miles from home, tears began to stream down my face. I heard an awful cry of pain and a high, shrill yelp, seeming to testify to my dog's execution. I screamed aloud, "Who is killing my dog?"

I stopped at the summit and called my mother from a roadside phone. She answered and I asked, "Who is killing my dog?"

She was crying, confirming it by her own tears. Fred's dog was there as well. I continued on to my destination, my anger turning into rage. At the bottom of the mountain, I noticed a dog identical to Ruffles, lying dead along the roadside. I stopped the car and ran back to see if it was Ruffles. Tears blurred my vision. The dead animal was identical. Confused and distraught, I turned back and ran alongside the dark, dangerous freeway to my car, leaving the animal on the road. Was it an apparition? I was determined to calm myself, so I continued on to find solitude and serenity at the temple's meditation that evening.

Earlier that afternoon, a nearby neighbor shot my brother's black Labrador in the tail. Fred was furious—threatening to shoot the neighbor. Another neighbor, an elderly lady and friend of my mother's, decided to put her large Airedale to sleep. That was her way to end the neighborhood dog crises and complaints about dogs. My mother had an Alaskan malamute that also paraded around the neighborhood. The animals' play dates were about to end abruptly. Unfortunately, the entire situation proved to be far too much for my mother to handle. My dog ran to her many times for his own salvation, food and comfort. But this time a family member took upon himself to end the chaos and shot my dog to death.

My dog's only natural enemy, it turned out, was man.

Our finances plunged into disarray, leaving us in a financial hell. This motivated Katrina to get a driver's license and find work. Our children suffered immensely during this time—we were always arguing and fighting. Art was his mother's best servant. Lesa and her boyfriend were arrested and thrown into jail for auto theft. Katrina came home about four o'clock one morning reeking of cologne and wine. I woke the kids and showed them the clock. I paraded around the house in a raging storm, and then cast the clock out through the dining room window.

The remainder of my attention was devoted to my dying father. His doctor had given him three months to live. It was my suggestion and agreed on by our family that I take him to see the psychic surgeons in the Philippines. I shared that precious time with him, during which he made a final attempt to turn old habits around. He shared with me some very unusual experiences and made me promise not to tell anyone. He had found peace with the God of his heart.

We returned home and my mother tended to his last days. She witnessed his final hour, an experience that will remain with her forever. In his last moments, he whispered his final special thoughts to her. His eyes were deep and hollow. His face and body movements relaxed then came to a slow stop. She noticed a disk of white light swirl just above his head. She followed this white brilliance with her eyes, tilting her teary gaze toward the ceiling. The brilliant light floated up and disappeared into space. At that moment of transcendence, a message of peace came upon her that things would be fine and not to worry about her future or the inevitable sting of death. I prayed for his soul's journey and protection. He imparted to me a message about the Egyptian Ankh that revealed a deep meaning of immortality. To this day the knowledge and wisdom he imparted to me—the exchange between father and son—will remain sacred and secret.

I took all my unsolved problems and laid them at the foot of my inner self. I changed my environment and began a new order of self-discipline. Each morning I would arise at four and take a quick shower, after which I would light three candles in the shape of a descending triangle. I would light special rose incense and begin to transcend my personal thoughts and prayers. I used vowel intonations and meditated for about thirty minutes each day. Within moments, I found myself rising above my worldly thoughts and worries. My intuition was my guide and influence. My prayers were intuitive, directed to a particular place, person in need, or situation. For my efforts, I received a golden light of peace that would slowly surround my whole being. After meditation I would write in my journal, describing my journey into cosmic consciousness.

Often Zephera would stand in the silent darkness with me. I would glance at her with a smile. Without a word spoken she would sit next to me, pick up a pencil, and begin drawing various shapes and symbols. An hour would sometimes pass with no words exchanged. Only our inner peace surrounded us. Our silence was broken only when the alarm clock would ring in Lesa's room. Zephera would then help me prepare breakfast and lunches before all the kids went to school.

Gradually, everything began to come together again as I slowly dug my way out of a deep, dark pit. One morning after breakfast, Katrina and I were discussing our divorce while Zephera happened to be listening in.

"Daddy, but who will take me to the temple?" she asked.

In March of 1981, after so many reconciliations, second and third chances and fractured communication, our marriage was truly coming to an end. I had stopped fighting to keep it and began a long fast with water. I needed to clear all my thoughts and create a new and wholesome atmosphere for myself.

One night, when I was well into my fast, I fell asleep on the living room couch. The evening was still and quiet. Katrina and the children had gone to sleep. As I was lying on my back, I began drifting out of my body. A comfortable coolness poured into me. I took enormous strides through space, accelerating through a great distance. I moved past many lights from large cities and eventually came into view of a stone pyramid as large as a mountain. My fog-like entity approached the stone structure just below midpoint on the north side of the pyramid. My faculties were keen and objective. I accelerated even faster toward the stone wall and went through it like a cool breeze, impervious to its opacity.

I moved into a small room near the base of the pyramid. My ghostlike body drifted upward through the rough-hewn corridor, at the end of which I entered another, larger room.

A beautiful woman with long, black hair approached me. She took me to another entrance and led me through a narrow pathway to the inside center of the pyramid—just to a certain point—then disappeared.

I stood alone in the growing darkness. Then, another woman approached me from a different direction and took me by the hand. Her joyful love and gentle softness was palpable as she led me to the very heart of the pyramid. She walked with me to the end of the path and stopped at the edge of a sharp cliff.

In front of me was a view of a vast, open sky. I could see the stars and all the wondrous workings of the universe. The distance was infinite. There was infinite space above me, in front of me, on both sides, and far below me. It was all one stellar infinite time and space.

Intuitively, I felt an urge to step off the cliff and into this vast, endless world before me. I looked directly above me and saw a shaft of liquid white light pouring down from the apex of the pyramid. This was my way out of total darkness, to ascend the ladder of white light and ride up through its peak. I now needed to overcome the infinite world of time and space and ascend beyond the illusionary body of the universe—but how?

It was humanly impossible. I hesitated for a moment and then managed to turn away from this inner challenge. I wanted to go back and be with the dark-haired woman who had led me to this place. I turned around and attempted to return.

When I approached her, she faded into total darkness. The more I attempted to reach her, the thicker the darkness grew.

Lost, I tried to return to the infinite cliff, hanging near the edge of the impossible. The only way out was to stand beneath the white shaft of light and ascend through the apex of the pyramid.

Suddenly, from out of the mist and behind the curtain of space and time, I heard a strong and bold voice. "Know that I am with you." I quivered with confidence and power as the voice resonated throughout my body-less spirit. Quickened by an unusual strength, I took a step forward and entered the bottomless pit of space. I moved through the veil of space beyond light speed. Time folded away, space evaporated into nothingness and disappeared.

Time did not exist. Space and time formed an illusionary curtain woven from the light of the stars into all the nebulous works of creation. I landed on the other side of the infinite world of space and time, standing on a stone platform. I again stood directly beneath the long shaft of white light. I was destined to travel up this shaft of light and through the opening at the peak of the pyramid.

Absorbed by an infinite power, I noticed a large meteor falling toward me from the apex—moving at a great speed. The closer it came, the larger and darker it became. Within seconds, it eclipsed the white light. It was like the moon falling on top of me or perhaps it was me traveling toward the moon at light speed.

I waited for the great voice to return with its mighty power, hoping that it would destroy this colossal obstruction that seemed destined to destroy me. The voice did not return. I remembered that the voice had awakened a confidence within me. I went into the core of my being and was consumed by a singular thought that quickened every fiber and cell in my body. I held this ineffable thought of power and let the monstrous mass fall to its destined collision course.

The instant the massive chunk of matter touched the top of my head, it exploded and disintegrated into thousands of pieces, spreading its debris within the pyramid strata of time and space expanding our finite universe and creating new forms of spiraling nebulae. At that moment, unscathed, I shot up through the shaft of white light. I swiftly moved through and beyond the peak of the pyramid into an eternal atmosphere of white light leaving the known universe behind.

The next morning, I went to the Rosicrucian Museum to do some research at the library. Upon my arrival, the first thing I saw was a life-sized statue centered in the middle of the pond at the front entrance of the museum. It was the first day of the display of the Goddess Taurt. This cold stone idol took on the embodiment of my memory of Katrina. This is where our memory froze in time; only a stone remained.

Taurt was originally a predynastic hippopotamus deity whose name meant "Great One." She was usually represented as a pregnant female hippopotamus with pendulous human breasts, the hindquarters of a lioness, and an elongated tail of a crocodile. In each hand she held the hieroglyphic sign SA, which means protection. As the "Great Mother," she assisted in the daily rebirth of the sun god, Ra.

Taurt was much revered throughout Egypt during all periods and at all levels of society. She was regarded as the protectress of women in pregnancy and childbirth. She was especially prominent in the eighteenth dynasty and is said to have assisted in the birth of Hatshepsut, one of the few woman who occupied the throne of Egypt.

Statues and amulets were placed in the tomb so that Taurt could protect and watch over the deceased during rebirth into the nether world. Taurt had temples at Thebes and at Deir-el-Bahri, the site of Hatshepsut's temple.

Taurt's great pain was the pain shared by women as a whole. I became filled with emotion and sorrow. I began to cry silently within. This was the end of my marriage to Katrina and, hopefully, the end of the pain that haunted our memory for so many lifetimes.

The museum attendants seemed proud of their new stone display. They spoke to a small crowd about Hatshepsut's ancient past. While others marveled at the display, I presided over the stone idol in silent mourning. I saw only the death of Katrina's memory, and soon I became free and loosened myself from its clutches. I finally let Katrina go from my heart, mind, and the past-life memories of her. Without me, Katrina now had to bear an enormous karmic burden.

My thoughts were on Katrina and the cross she was bearing.

Saddened and tearful, I turned from her past and left the ancient shadow behind me. The world's sorrow encircled me. I left humble in faith and the knowledge that God brings life to the thousands I heard crying His name. In this house of sorrow and the shadow of pain the veil of death remains. It was through the return of God's presence that all their ways could be healed and their sorrows removed. The world's sadness and sorrows were upon me. I strolled quietly and in full humility to the Rosicrucian Library to research the deeper meaning of the Great Sphinx Tablet.

The legends of the ancient mystery schools of Egypt tell about the sacred obligations undertaken by candidates for initiation between the paws of the Sphinx. It was symbolic of temporal and divine power. Completing my research, I left the library and walked past the roses that filled the air with an all-pervading aroma. I sat on a bench next to the Ankhaton Shrine and noticed several staff members carrying electronic equipment to the research lab.

I recognized one of them and shouted a sarcastic reminder in jest. "Make sure the camels have plenty of water." I realized I was treating these great researchers like second-class citizens, but I could not seem to help myself. They looked back in a puzzled way but remained unaffected by my words.

My comment was really about my own long journey home. It was about me crossing a vast desert of an ancient memory and knowledge. I arose from the bench, went to my car, and started the drive home. On the freeway, the traffic was heavy. An emotional charge began to form deep within me, surfacing into a prayer. Is there no answer for civilization? Have all nations trodden the path of the fool? Are we about to be another fallen world society?

Tears from the depths of my soul began to stream down my face. Must the culture of man end? Must I witness again an age of knowledge sink into the shadows of death? Was there no way out from the monster we had created and become? Suddenly, a burst of energy pushed down through the top of my head.

I felt like a modern-day Atlas with an unbearable weight upon my head and shoulders. It was the sum total of all negativity in the civilized world. God was my witness. I cried out loud to the Creator, saying,

"With one quick glance from your eternal eye, let us again become a great world with a knowledge that replenishes the crude mechanisms that mankind depends upon. Must civilization become just a dim vision for the immediate and future generations? Please help me, God, before man assumes the world to be a material existence!"

Katrina would never understand me. I could not be sure if anyone would, but God did. As I approached the Santa Cruz Mountains and made my way up to the highway's summit, my consciousness was lifted simultaneously. Again I was at one with all things and everyone around me. My awareness was filled with a new vision and a new beginning. As I approached the mountaintop, I focused on the power of the white light shining on everything and within everyone. The sun spread its light equally over the Earth; so did the consciousness of mankind for those who thirst for it.

I understood the vision as it was recorded in Genesis. It was a birth of a new world consciousness. It was only a small glimpse of the memory of what once was upon the Earth. This was a vision of eternity and God's promise to keep his fallen creation in motion.

My questions were answered as I witnessed the light robe the planet with the Creator's eternal mantle of compassion and promise to mankind. The Earth was protected like a mother who swaddles her child to protect it from outside forces that may do harm. I was comforted. The Creator blew light upon a planet that was fast approaching destruction. I arrived at the peak of the mountaintop and was filled with a light that enabled me to understand all the needs of humanity. A grace and a new era of diplomacy fell upon me as I witnessed the light mend the ways of all mankind.

After driving several miles, I noticed a young man, hitchhiking. He was about seventeen years old. No one would stop for him; the throng of traffic was too thick or people were just too frightened to help him out. I became annoyed with the people around me. They were like blind worms pushing their way through their own dung heap. Suddenly, my thoughts shifted to Henry Kissinger. To me, he symbolized an archetypical world. An economic king, whose only desire was to dominate human destiny. I believed he was preoccupied with unconscionable and obsessive acts against humanity. He's a diplomatic prophet who unknowingly serves the will to the god of this world.

I questioned the Creator, asking, "What am I, if nothing more than your will of creation?"

With this brewing in my mind, I stopped and pick up the hitchhiker. As he entered my car, I was silent for a moment. Somehow, I knew him. Somehow he was reading my thoughts. I turned to him and said, "Henry Kissinger is ascending towards one world government power!"

He smiled at me and remained silent.

My consciousness was making an unusual descent, during which I began to view my life as a past memory. I stood in my own past, seeing my future being carved into fetuses from one lifetime to another.

I was observing my present self from a future life, while looking back where I had once lived many years ago. I made my descent below the summit of the mountain and into the past memory of myself. I was viewing my past and at the same time drawing near to my new life.

I was no longer Dan Weiss, but someone reborn into a future lifetime. My future was a product of my own creation and experience. A secret was revealed to me: it is our memory that incarnates into another so-called lifetime. I saw my next incarnation carving its way into the future.

My soul went into flight by the life-giving spirit of God. It was this cosmic circuit that made it possible to return to God by way of the ascension. This all took place in a matter of minutes.

Finally, the hitchhiker broke the silence, "My God, Jehovah, has come in 1976!"

I looked into his piercing blue eyes and quietly said to myself, Praise the Creator and may our debt be remembered no more! And nothing more was said. I let him off in Santa Cruz. An aura of freedom accompanied my companion as he left the car.

I drove through town, bound by memory only and the people I have known.

My thoughts soared far into a future life when civilization as we knew it had all changed. The people I knew in the life of Dan Weiss were fast becoming a distant memory. Once again, I was strolling through a vast open valley of death.

I drove to the home of my brother, David. He was like an earthly king to me. Yet his personal destiny was in opposition to a new-age future.

As I drove, my mind was caught up in David's trials. I could see that he was trapped in a battle between man's destiny and God's. It was the age-old struggle between the forces of light and darkness and the destiny of such forces. I had to overcome his inner rage of cause and effect. I was at one with the universal mind, and my every thought and action was filtered through that mind. I met his internal opposition and a battle commenced within this universal mind.

An intense friction began to grow between us. A vision of war was being created, to be played out on Earth in a bloody nightmare of an actual battle. He could not understand my approach. Two opposing thoughts between brothers can spawn a raging battle and spread into a contagious war anywhere on earth.

All wars are created in the mind and fought on a battlefield. All wars need to be fought in the mind and end in the heart. Opposing forces create wars, but if they are fought in the heart, mankind finds compassion and forgiveness.

Our consciousness became fused like an atomic bomb with the added strength of many nations boiling under a real threat of war. Our causal thoughts created many opposing forces of war upon the Earth. I began to direct our warlike entity like chess pieces on Earth's playing field. We were like two great opposing Gods displaying our differences, playing them out on a real battlefield. Our opposition was so great that either war or wisdom would prevail within the hearts of men; only the lack of wisdom would lead to world crises.

These thoughts attempted to transcend our disparity. I made a decree out loud to neutralize this great opposition to war. I called out to the consciousness of the free world and decreed, "I am Denmark!" I invoked a higher, conscious form of intellect and liberty, democracy and independence. Instantly in living color, I could see the country and its cities in my mind.

"I am Sweden!" I exclaimed out loud, again as a declaration to neutralize and counteract David's opposing inner forces of planetary war.

My consciousness rose as a spiraling white fire and touched a world of absolute nothingness. The circumference of this consciousness expanded so much it literally swallowed our cause of war upon the Earth. The cause of war was beginning to dissipate from the depths of our minds. Now I was destined to overcome the conscious battles between brothers.

An equal force of this power opposed me as I approached David's driveway. I wanted to speak to him most urgently. I knocked on his front door and entered at the sound of his voice. He was seated quietly at the kitchen table; I sat across from him. He was busy with some paperwork.

I tried to explain my personal experience and understanding of the Book of Revelation to him. With a stern voice, he said that he was not interested in my Book of Revelation. He was not paying any attention to my subject matter or me. He left the table and went into another room. His lack of interest frustrated me.

The phone rang and I heard David talking to our mother.

"Yeah, I know, Mom. That's what I'm just trying to tell him. He's on some sort of weirded-out Vietnam trip again. I don't have time for it."

Then I spoke to her myself, telling her, "I am standing for a cause of faith that God kindled in my heart." Once again, the entity of death slowly began to surround my body. God filled my mouth with words to speak that would keep those I love close to me. I assured her that we one day would know our mission with the Creator.

I hung up the phone and went back to talk with David. He said my eyes were dilated and asked if I was all right. His skeptical attitude suggested I was on drugs. He didn't understand anything I said, so I didn't try to explain in any detail what had occurred that day. I told him I was extremely tired. He didn't ask me to share anything. I wanted to talk about how the Creator was going to swallow death; instead, I swallowed my own thoughts and left silently.

Floating through the shadows of death, I knew for certain that humanity was aboard a ghost ship caught up in a torrential storm of ignorance, disease, poverty and death.

I left David's home without a goodbye. I drove to my parents' home, thinking about the hitchhiker, when these words from Isaiah echoed in my head:

Behold the days come, saith Jehovah, that I will make a new covenant with the house of Israel and with the house of Judah: not according to the covenant that I made with their fathers in the day of my taking them by the hand to lead them out of the land of Egypt; which my covenant they broke, although I was a husband unto them, saith Jehovah. But this is the covenant that I will make with the House of Israel, after those days, saith Jehovah: I will put my law in their inward parts, and in their hearts will I write it; and I will be their God, and they shall be my people. And they shall teach no more every man his neighbor, and every man his brother, saying, 'Know Jehovah; for they shall all know me, from the least of them unto the greatest of them, saith Jehovah: for I will forgive their iniquity, and their sin will I remember no more.

One year later, everything started to change for the better. With a ticket for Iceland and final divorce papers in hand, I had finally dug my way out of the ruins.

Chapter Twenty

~ Ultima Thule ~

I am everything that was (the past) and everything that shall be (the future), and no mortal has ever removed my veil.

—*Freya, wife of Odin, the all-father who dwells in Agartha, the center of the earth*

"The Ultima Thule is known as the land farthest north. A legend tells of a paradisiacal land of great beauty and perpetual light, where there is neither darkness nor too bright a sun." My imaginary listeners awaited my next words.

"This wonderful land is described in Halldor Laxness's Nobel Prize-winning book, *Christianity at Glacier*. This is where people are different because of the mystical power of the Snaefellsjokull glacier, where they have lived in peace and tranquility at its foot for centuries. The gateway to another world was there in Jules Verne's Journey to the Center of the Earth. One can scarcely differentiate between the mythological and the real that pervades these vast Nordic areas. In the vast stretches of the far north, large lakes may freeze, but tropical animals roam in herds, and birds of many colors throng the sky. There, it is a land of perpetual youth, where people live for thousands of years in peace and happiness." I ponder my image in the mirror before continuing.

"Yes, to live for thousands of years in peace and happiness. Just think of it. I remember a journey to such a place, once. I was a child when I saw the movie version of Journey to the Center of the Earth with Pat Boone and James Mason. Perhaps, just like me, you were so inspired by that movie that you have always wanted to travel there yourself, to that 'land beyond the Pole,' as Admiral Byrd called it. Perhaps your memory, just like mine, was seeded with a glorious dream similar to the one Jules Verne described as his party of explorers entered a volcanic shaft. And after traveling for months, they finally came to the hollow center of the Earth itself, experiencing this new world with its own sun, oceans, land, and even cities of Atlantian origin. Perhaps you, too, had this same glorious dream, which has never faded. Perhaps that is why you are here today with me in Washington, DC.

"Did you know that in the early nineteenth century, the belief that the Earth was hollow and habitable—with entry to the inside world somewhere in the Antarctic—was so popular that Congressman Jeremiah N. Reynolds was able to introduce twenty bills into Congress, proposing that a navy expedition sail into the globe's interior? The issue continues to be a topic of serious scientific discussion.

"In his work entitled Moongate, William L. Brian II cites a scientist named Joseph Goodavage, who spoke at the 1961 World Earthquake Conference in Helsinki. At that conference, Goodavage talked about a remarkable phenomenon recorded during the May 22, 1960, earthquake in Chile. What was recorded, he says, was the sound of great ringing. The Earth was literally ringing like a giant bell. The Earth continued this ringing sound for a considerable length of time.

"A bell, ladies and gentlemen, rings because it is hollow. This, I give you to consider as possible evidence that the Earth might indeed be hollow, and that we might indeed actually be able to go inside. Ladies and gentlemen, this attempted expedition is long overdue. It is now time for us to take this journey together."

I paused, scanning my imaginary fellow adventurers who were prepared to fly north of the 80th latitude this very day. I shuffled my notes. I know I should tell them about the Iceland trips. I found the right place in one of the many file folders scattered across the shelf in front of the bathroom mirror.

The telephone rang. "Yeah, Dan here."

It was Wanda. "Danny, what's up?"

"Hi Wanda, I was just putting together my orientation lecture for the trip. I should paint the whole picture for them. You know—how I got there, whom I met, what went on, the whole ball of wax."

"Orientation lecture? You mean, like school?"

"No, of course not, not that kind of lecture, I mean just to get them mentally, psychologically, and spiritually prepared for the whole thing."

"Spiritually prepared? Jesus, Dan, from what you've already told me, that'll take years! Besides, if they're going on this trip, they're already into that sort of thing. No one wants to sit through hours and hours about someone else's out-of-body UFO stuff. I mean, is that all you're going to do? Is this all a big hoax and you're not really going up there at all?" Wanda was starting to whine.

"Of course not. We're really going. I just want them to have the whole picture first. I want them to understand how I personally fit into this and what's driving me to find out the truth. Is the Earth hollow, or isn't it? Is it our lost Garden of Eden, or isn't it? Is it hell where the Devil lives? Where's the truth? That's the whole point here, Wanda."

"Okay, like what?"

"What like what?"

"Like what are you going to tell them?"

"Oh, okay. Like who I met before I went up there first, and about that lady who did those predictions that all came true, and how I met Gutha and that whole thing about her investigation, and..."

"What investigation?" Wanda sounded interested. "Maybe you'd better start at the beginning." Her voice was now calm, instructive.

"Start at the beginning? Haven't you been listening to anything I've said over the past few months?" I sputtered.

"You know what I mean—start with your first trip to Iceland."

So I began from March of 1982.

"I was making preparations for my trip to Iceland. I called the Rosicrucian Order in San Jose to see if anyone there could put me in contact with a member in Iceland. As it happened, the man in their tape and video department was Icelandic and they gave me his number.

"I called him up and we got together for lunch. Not only did he give me names and numbers of people to see when I got there, but he also told me about the California Icelandic Society. He invited me to the Icelandic celebration, Thorrablot. It was a lavish dinner all flown directly from Iceland. It's a major celebration in honor of the mythological god, Thor. He said the president of Iceland would be there as well as the queen of Denmark and other interesting people I should meet."

Zephera wore a fine white dress and I wore a light gray suit with a burgundy tie.

We met nearly everyone in a quick passing. I was introduced to the pilot of Icelandic Air and his wife, who taught Icelandic at Stanford University and who later became my language instructor. I was given several contact names and an offer of an apartment in Reykjavik, once I got there.

We got a real taste of a modern Viking celebration. The party lasted well into the late hours. A highly enthusiastic group of people sang folk music, danced, and dined sumptuously on everything from smoked lamb, head cheese, cured shark, and tripe rolls to blood pudding. It was a magnificent experience.

One of the guests, Sveinbjorn Beinteinsson, became the topic of conversation over dinner. Sveinbjorn was the high priest of the recently revived sect called Asatru—the religion of pre-Christian Iceland. He was also the country's leading expert on traditional narrative poetry and storytelling, Rimur. He would be in our greeting party at the airport in Reykjavik, along with Gutha. And Gutha introduced me to Halldor Laxness, whom I hoped would be at the airport too.

Sveinbjorn Beinteinsson

"You sure met all the right people," said Wanda.

"Not only that," I continued, "every year at this celebration they have a drawing for a round-trip ticket to Iceland, and guess who won the prize out of nearly four hundred people? Zephera!"

"That's remarkable," said Wanda, truly amazed.

All this excitement before my journey only served to make me more anxious than ever, I told Wanda. One night while at a Rosicrucian Convocation meditation in San Jose, I noticed a woman whom I did not know observing me from across the table. After the meeting, she approached me and said, "I sense you are traveling north."

"But how far north?" I asked.

"Farther than you might think," was her reply.

"Oh really," I commented without emotion, wondering just how much this woman knew, and how she knew it. "And just what else do you 'sense' about me, Ms—?"

"Stubbs," she finished her name for me.

"Ms. Stubbs, it's great to make your acquaintance—and what else?" I was curious by now.

"Perhaps we could get together next week," she said, "and I could do an entire reading about everything I am seeing."

"Oh, really? And do you 'see' this kind of thing often?"

"As it happens, Mr.—?"

"Weiss," I finished my name for her.

"Mr. Weiss, I am a psychic."

"Why doesn't that surprise me, Ms. Stubbs?" I laughed. "How's next Wednesday afternoon? I believe I'm free then."

"Perfect." Ms. Stubbs was agreeable.

"Wednesday afternoon it is, then." We agreed on a meeting place and time. "I'll be there, Ms. Stubbs." I kissed her hand as we parted. The following Wednesday, we met for more than an hour, during which time she made three specific predictions. All came true within that year.

"While I'd had no experiences up until that time," I told Wanda, "nor was I interested in UFOs, she accurately predicted that I would have a major UFO encounter and make an unusual discovery, and also that my daughter, Zephera, would go with me to Iceland. All three predictions seemed either off-point, or completely impossible. As it turned out, however, all came to pass—the first two were just thirty days before my departure.

"The UFO experience was accompanied by the voice of Ritter von X in my mind saying clearly, 'That's good. You've been beamed aboard the craft but cannot understand it, yet. In time, you will, Danny. In time, you will.' "

Wanda's voice on the other end of the line interrupted me. "Well, let's hear it. The UFO experience—let's hear it!"

"I'm getting to that, Wanda. Be patient. I was awake, just lying on the living room couch. I was literally taken—carried away—by several beautiful golden lights, golden spheres that were as large as grapefruits. They entered through the walls of my house and moved across the room, approaching me. I was awestruck by their beauty and magnificent presence. There were ten golden spheres joined together but somehow not touching each other. These 'dancing lights' moved directly at me, swirled above my head, and made their omnipotent descent by pouring through the top of my head accompanied by a sound comparable to a 747 jet on takeoff. The lights slowly descended into my whole body and soul. For a few moments, the dancing lights swirled around within me. The swirling continued with an upward spiral movement through my body—swirling in, through, and around me. They suddenly took me. I was simply gone.

"I remember nothing. But I had been taken somewhere. For how long, I am not sure at all. And to where, was completely unknown to me. I don't know if I was gone for an hour, the whole evening, or several evenings. Time did pass, but I was not aware of how long. I was then returned the same way by the dancing golden lights, as when they first entered my room. I once again became conscious as the dazzling golden brilliance encircled my head. The lights entered with the same familiar sound of a 747 preparing for takeoff. All at once they raced up and through the middle of my conscious self and exited through the top of my head. I was then released into my full waking faculties. I was fully and consciously awake as I had been when I first witnessed their ineffable beauty. The prediction about Zephera came to pass soon afterward."

"How did that ever happen?" Wanda interjected. "I thought Katrina would never let you have her."

"That's what I had thought, too. Then one day she phoned. She had been remarried for about a year. Five days before I was to leave for Iceland, she called up to say her new husband was beating on her and she couldn't take care of Zephera anymore. She wanted me to have her because I'm the only one she could trust."

"Gees, that's a switch. It sounds kind of manipulative to me," Wanda said.

"Yeah, that's what Art thought, too. He told me, 'Hey Dad, if you go back with her I'm out of here on the first train to anywhere.' He was more than ready for a new adventure, just as I was. Art was going to Iceland with me, too. I knew he didn't want things to get screwed up again between his mother and me. He had seen too much of my ping-ponging with Katrina over the past ten years. He knew that I was perfectly capable of running back to help her at any time. I told Art not to worry. I would just go and get Zephera and leave and that would be that. She would go to Iceland with us. And that's the way it came down."

"Are you going to tell them all that?" Wanda's tone implied criticism.

"Tell who?"

"I mean tell the people who are going on this trip—the participants and are you going to tell them all this in your orientation lecture?"

"No. Yes. Well, yeah, part of it, anyway. They don't need to know all of it."

We laughed together.

"So what else are you going to say?"

"The stuff about the Icelandic legends and how they seem to parallel a lot of other ancient writings."

"Yeah, like what?"

"Like the Koran for one. This is really neat, Wanda. Remember that forty-day fast I did in 1989? Well, during that particular fast, I read the whole Koran!"

"No! The entire Koran?" Wanda was skeptical.

"Yes, I read the whole thing, Wanda. And, there are references throughout about the center of the Earth. I have them all highlighted—documented—everything. It's really uncanny how these myths, legends, the Bible, and other sources, some of which I can't divulge, you know—how they all speak about the same thing and Byrd too. Remember? He and his crew confirmed it. It's all in that secret diary."

"I know, you've told me all that before. What about the Koran? What does it say—exactly?" Wanda paused. "I'd better go turn off the iron. Call me back in a minute, Dan." Wanda hung up.

What? She had left the iron on all this time!

Thinking about the Koran, I reached into my file folders, looking for the research. I got sidetracked in the Iceland file and began to read aloud to myself.

"It is well-known and written in the Old Icelandic legends that Odin was the sky god and the All Father. He is known by his blue cloak, gray garments, one eye, and long, white beard."

One eye? Yes. That's it.

Fumbling again through the files, I found what I was originally looking for. I continued to read from my notes about the Koran.

"Koran, sometimes spelled Quran is the Holy Book of Islam, is regarded by Muslims as the true word of God and the only true religion that was revealed to the prophet Muhammad and collected in book form after his death."

I continue to scan the Koran file. Finally, I find the reference I am looking for, and read it aloud. "The end of the world will be announced by the coming of the Mahdi. This means the directed or guided one. The Mahdi will slay the Daijal, the one-eyed evil spirit, and combat the dangerous enemies, Yajuj and Majuj, or Gog and Magog who will come from the north of the earth."

Dr. Stranges and Ritter von X would surely have different interpretations of all this, I thought to myself before continuing.

"The dreadful angels of hell and the horrors of that place are as thoroughly described by the theologians as the pleasures of Paradise, with its waters and gardens underneath and the houris who are the virgins. The Houris - pronounced hu-ree, means the beautiful maidens who, according to Muslim belief, live with the blessed in Paradise."

I stopped reading, picked up the telephone, and dialed Wanda's number.

"Wanda, did I ever tell you about my first trip to Iceland in 1982?"

Wanda was on a single track. "What about the Koran?" she prodded.

"This is more important now. It was June sixth, to be exact. We landed in Reykjavik, and were greeted by the Rosicrucian Master of Iceland, a fellow named Denni. We had arrived two weeks earlier than they expected. We moved into a vacant apartment for a while, until the one that the president of the Icelandic Society had arranged for us became available. If it hadn't been for that change in plan, I would never have met Sigurlena."

Lena. What a magical gift she was. I was in Thors Café one night. I saw her from across the room and approached her. We talked and danced most of the night. I had a great time and then asked if I could walk her home.

"Walk me home?" She was visibly surprised.

"Yeah, why not? All good American gentlemen would ask."

"Oh, I could not do that," Lena replied, dropping her beautiful eyes.

"Oh, of course you can," I gently touched her cheek and smiled.

"Well, maybe it will be okay."

It was not until later she told me about an Icelandic tradition that if a fellow offers to walk a girl home and if she accepts the offer, that means she is extending an invitation to spend the night with her. Sometimes ignorance can get you a long way.

Lena—what a gift, I even wrote a story for her called, "The Magic Gift."

And then there was Gudbjorg Gudmunsdottir. Her nickname was Gutha, for short. Her name means the right hand of God. One of my Icelandic Society contacts gave me her phone number. I called her up one day and invited her to meet with me and Zephera.

The night before we visited Gutha, I was sleeping in the daylight (it was daylight most of the day and night there), when I had a beautiful dream about ascending through golden clouds while reciting Robert Frost's poem, "Fire and Ice."

To get to Gutha's apartment, we had to walk to the top of a tower filled with many little square windows. The sun was shining through the clouds outside and streaming through the windows. The clouds created a golden mist throughout the entire castle-like tower. It was exactly like the dream.

"Gutha lived at the very top overlooking the capital city, Reykjavik,"

I recounted to Wanda. "She had two beautiful daughters, Hebba and Agusta. They were a few years younger than Zephera. Gutha was an airline stewardess with Icelandic Air. This kept her out of town for a week or so at a time, and her husband was currently away in San Francisco for the next three months researching a NASA project regarding moon rocks. She needed help. We worked out a plan. Zephera's job was to help babysit and get her children to the school bus stop. My job was to cook and clean, in exchange for free rent. We stayed the rest of the month in the first apartment, and then moved into Gutha's for the next three months. I was really beginning to like this place."

"Are these women both gorgeous?" Wanda did not really want to know the truth.

"Wanda, they're Icelandic. Of course they're gorgeous!"

"Did I have to ask that? What did her husband have to say about the whole setup?" Wanda was not sure she wanted to know this, either.

"Wanda, they're Icelandic," I repeated. "The men and woman are totally independent of each other, yet empowered by a love that comes not from this world. That's why all those countries up there don't go around starting wars."

"Well, that's one thing they've got going for them. That's why they had that '86 Summit there, probably. You were there then, too, weren't you?"

"1986? Yeah, Zephera and I went up there then, too. Wow, was that ever weird."

"Why? What happened?"

I recall 1986, as Wanda listened.

The minute we got off the plane, Gutha and her brother-in-law met me. "Get in the car. We need to talk." I knew right away something strange was up. "I was investigated by some men and so were my parents and friends!" Gutha whispered.

"Investigated? What do you mean?" I did not understand what Gutha was talking about.

"Right after you left last time—two years ago. These men came to the door and started asking all kinds of questions—how did I know you? Where did you live? What things did you leave behind? How did you get here, how did you leave, what airline, what plane? It was like they thought you sneaked into the country illegally—they didn't know how you got here. They started pushing things around. I got scared and called my brother-in-law. That's why he's with me here, too." Gutha handed the inquiry over to her brother-in-law.

"I didn't know what was going on," said the brother-in-law. "I don't even know who you are, really." He was suspicious.

"What are you both talking about?" I was surprised.

"Dad, are we going to meet the president?" Zephera asked quietly.

I ignored her. "Take this whole thing from the beginning, Gutha. Step by step. Who came to your door?"

"Dan, I do not know," Gutha repeated. "They said they were from international security."

"Did they show you any identification?"

"Dad, are we going to meet the president?" Zephera repeated patiently, but I was listening intently to Gutha.

"Probably—I don't remember. They just pushed their way in." Gutha was trying to cooperate with me.

"Was it because of the letter?" I asked.

"What letter?"

"Dad?"

"Not now, honey. Maybe it is about the letter from Ritter von X regarding specific instructions. The letter I hand-carried directly to make contact here in Reykjavik, remember? I showed you. That's why I came back in 1984, just for those thirty days. You wanted me to stay longer, remember?"

"I remember you were here for a short time, that's all. And then those people came. Danny, what's going on? Were they after that alien geode you showed me?"

My mind raced to July 19, 1984, when Ritter von X sent what he called a "four-point letter of the utmost importance." I'd shown Gutha the letter, which told of unseen world events and Ritter von X had been authorized to contact me about presenting that letter, which I had done. He had also told me about the crystalline generator, which protected the Golden Age from anti-matter, and he'd made an obscure reference to alien spaceships pictured above the castles in Cranach's painting. I was instructed to burn the letter, but not before Gutha had read it.

Otherwise, its message was kept secret. My question to her was, "Is this about your ancient Icelandic heritage and legend?"

"Not mine," she said. But who had investigated her, and why?

Gutha, her brother-in-law, Zephera, and I sat quietly in the car for a few minutes before he started the engine.

"Dad, will we get to meet the president?" Zephera asked once more, taking advantage of the silence and interpreting it as hopeful.

"Reagan?"

"No, I mean Gorbachev, Dad. Will I get to meet President Gorbachev?"

"I don't know, honey, we'll see. Everything will be okay, don't worry." I tried to be gentle and consoling as we drove to Gutha's castle tower in the golden skies of Reykjavik.

Later that day I answered a knock at the door of our room and looked into what appeared to be ancient eyes resting beneath wild blonde brows of a woman in her fifties. She wore a headscarf, a dirty printed apron, and blue jeans. She was carrying a camera.

"This seems to be an ideal spot for a few photos, sir. I wonder if I could just—" She stepped past me on her way through into the bedroom. "Maybe take some pictures from here of the summit. Both presidents will be coming right down that street, you know." She pointed to the school across the street, which had been swiftly transformed the day before into a high-tech satellite communications center.

Zephera and I looked at each other in amazement.

"Excuse me?" I followed the woman into the kitchen.

"Photos, you know, to remember the big day. This is an important summit, sir, in case you don't know." The woman headed for the bathroom, peered around suspiciously, and then looked out the small oval window onto the street below.

She seemed to have simply invited herself in. She strode past me, ignoring me, so I said, "Yes, sure be my guest. Come on in anytime. The door's never locked."

As the woman left, Zephera and I again looked at each other questioningly. "Oh well. This is Iceland," I said as I shut the door behind her.

"Did she ever come back to take the photos?" Wanda asks.

"No, never."

"Aha!" said Wanda. "So then what happens?"

"Aha, indeed. Well, one of our contacts named Rosa invited Zephera and I for dinner, and later invited us to the American Embassy party with Reagan's secret service guys. Reagan was to be there, too, and perhaps Vigdis, the president of Iceland."

"That's pretty neat. Did you meet anyone important?"

"I met the prime minister of Iceland and other interesting people, but I felt like a celebrity myself. Zephera was all dressed up in her best white dress, and I was in my usual gray suit and burgundy tie. Getha, another beautiful blonde, took me by the arm and escorted us through the security gates and into the grand dining room, where we were seated at the dinner table right alongside all the other important people. Zephera was fabulous. She answered all kinds of questions about politics and the environment, whales, and everything. She didn't hold back any of her own personal opinions. And she was very well-spoken."

"Oh yeah, that Greenpeace thing was happening then, too, wasn't it?" That's when some environmental fanatic sunk an Icelandic whaling ship, Wanda remembered.

"Right! And maybe that's why they were following me everywhere. I was a munitions and weapons expert in the military, remember? Maybe they thought I was blowing up the whaler."

"Dan, you know that's not why they were watching you." Wanda sounded dead serious.

"What do you mean?" My question was also serious.

A pause in our conversation allowed us both to think a while. Wanda broke the silence.

"Well, maybe I can understand the investigation on Gutha, but why the old lady with the camera in your apartment?"

"A perfect spot for an assassination attempt, I'd say." I was fairly certain about the reason why the older woman had barged into my apartment. "Maybe it was just all about security precautions."

"Same with the Embassy party, probably," speculated Wanda. "At least there, they could all keep an eye on you and not have to miss their own party."

"I can't blame them there. It was a great party. Later that evening,

Rosa and Getha invited us along with a small group of people to meet with President Gorbachev. He was presenting his world nuclear disarmament plan. Rosa suggested that I get a press pass from my hometown newspaper. I wired Mom to call our local newspaper owner and let them know where I was and what I was up to. They faxed me a pass right away and Rosa had it approved by the foreign minister. They issued me a permanent, laminated one especially for the Reagan/Gorbachev summit. It's in English and Russian with both flags. My number was 2494." What a memento!" Wanda was impressed. "I keep it hanging over my desk. That press pass got me to meet many interesting people and places. The meeting was right across the street from our apartment. It was being televised all across Europe and Russia. Zephera and I went over there with our camera and press pass and we heard the whole Gorbachev plan. We were no more than ten feet from where he was speaking. Zephera was on camera. She looked over to one of the monitors and saw herself. She got real excited and listened to every word he said. He had charts and timelines and everything to explain his plan for a nuclear-free world by the year 2000—in detail, I mean, right down to the last nation. The trouble was the American press spoke only about Reagan's Star Wars plans to all the American presses. Reagan spoke to America and Gorbachev spoke to Russia and Europe. What a sham. Nothing was broadcast about Gorbachev's master world nuclear disarmament plan to the United States. It was a great plan. At that time the world had stopped all nuclear testing. The two world powers could have made the world free of all nuclear weapons, but unfortunately, none of that happened. This was Gorbachev's proposal; Reagan's proposal was the Star Wars defense plan.

"The motion picture movie, *Top Gun*, was playing next to the hotel where the president of Russia was staying. All I could think of was, Dear God, here we are once again, the archetypical opposing forces playing war games on the chess board of planet earth. I believed Gorbachev's offer was genuine and the world's last chance to disarm nuclear weapons."

"Now the future of nuclear weapons will spread into many hands to all the wrong people for all the wrong reasons worldwide," Wanda sighed with disappointment.

"Listen, Wanda. I could talk to you endlessly about this, but I'd better get back to my orientation lecture. Talk to you later."

"Hang in there, Danny boy."

Again in front of the bathroom mirror, I rehearsed another version of my opening statement: "Ladies and gentlemen, as the duly appointed director of the International Society for a Complete Earth, it is my pleasure to welcome you here to Washington, DC, this morning. You've all read the reports and done your own research—that's why you're here. You all know about Admiral Richard Evelyn Byrd's expeditions to both poles. You all know of the controversies concerning Mr. Giannini's alleged radio message of "Godspeed" to Byrd at his 'Arctic' base in February of 1947, and the subsequent discrediting of that transmission on the basis that it was a typographical error and should have said 'Antarctic.' You are all aware, I presume that some still believe Admiral Richard E. Byrd was actually diverted and sent on a secret mission to the North Pole during the same time period of February 1947. I myself do not know what is true, ladies and gentlemen. And the truth is what I propose to deduce. I only know what my research has borne out.

"Observe, if you will, these overhead transparencies taken from National Geographic Magazine of October 1947. The entire article, from page 430 to page 522, is a monumental document, to say the least, about Byrd's Antarctica expedition. This is not documentation of the North Pole. However, many of the photos prove beyond any shadow of a doubt that the poles are not simply white, cold, ice and snow. And that huge, fresh-water lakes and ice-free oases abound. I propose to show you just a few of the most remarkable pictures; the rest you may read for yourselves.

"One such photograph can be seen on page 475, Plate VIII. The text says: 'Lt. Comdr. David E. Bungers...found them in a landscape that made them question their own eyes—a land of blue and green lakes and brown hills in an otherwise limitless expanse of ice. This oasis was by far the most important, so far as public interest was concerned, single geographical discovery of the expedition. The area of this ice-free region is somewhat more than 300 square miles.' "

South Pole Oasis—For the first time in history, Lt. Commander David E. Bungers landed on a blue-green, warm-water lake in Antarctica on a Martin Mariner seaplane in the midst of a remarkable 300 square miles of "oasis."

—National Geographic Magazine—October 1947

"Another, on page 516, shows men landing on a lake in Antarctica for the first time in history. The text below the photograph reads: 'Furrowing the blue-green water is one of the wing floats of the Martin Mariner seaplane which alighted in the midst of the remarkable "oasis." Beyond, rise icebergs and bare, brown hills where a superficial survey failed to disclose any visible sign of life.' Life, however, may indeed exist. Not on the land but inside. We shall soon see for ourselves.

"I trust you are all rested and ready to embark upon what will most likely be a historical journey followed by the entire world—the reenactment of Admiral Richard Evelyn Byrd's exploration to the North Pole. Indeed, we hope, our journey will also be the reenactment of Admiral Byrd's entry into the very center of the Earth itself, where we will most assuredly encounter adventures beyond belief—perhaps, ladies and gentlemen, even Paradise itself."

I placed my file folder down on the bathroom counter and gazed long into my own reflection. My mind wandered into the past, where

South Pole Plane—For the first time in history, men land on a blue-green fresh water lake in Antartica. In this photo, we see the wing tip of the Martin Mariner seaplane as it lands in a remarkable "Oasis."

—*National Geographic Magazine, October 1947*

I remembered an out-of-body experience that began with a very loud cat's meow.

It was about four in the morning when I was awakened by a loud screeching. I stood up and stepped out of my body, walking back into a memory of a former life. It was in fact the most recent life prior to this one. I approached the window and moved the curtains to one side. Outside, I saw a rather tall individual, dressed in black with a Tibetan appearance. He was familiar. We saluted each other in military fashion. I spoke to him in German. I began to feel his powerful influx of energy move up through my hands and arms, and then throughout my whole body, which was vibrating with this apparently superior terrestrial force. Our thoughts merged.

He invited me to join him and led me to a local power generating plant. I recognized the area immediately. He communicated telepathically, showing me how to increase my energy by directing my thoughts and touching the power conductors. I followed his instructions and began to receive an enormous influx of energy.

I asked him where the best entrance could be found to the inner realm. Suddenly, I found myself facing this individual near a huge, snowcapped mountain. I asked him for directions to the center of the Earth. He was anxious to show me the direction, and again spoke telepathically. I followed him, and as we began to climb the mountain, snow started to fall. Eventually, we came to a very steep, almost vertical rock cliff. But here I hesitated, telling my companion that my daughter still remained in Iceland and I needed to wait until a later time to complete our journey. He assured me that we would resume our journey only when I was ready and that he would wait patiently for me until such time. At that moment we stopped and braced ourselves against the vertical mountain cliff. He understood my need to stop. He would wait patiently until I was ready to continue the journey. He knew I had some unfinished business to complete.

The telephone rang. It was Wanda again. "Danny-boy, just thought I'd better check up on you to make sure you haven't checked out of the planet or something."

"As a matter of fact, I was just thinking the same thing, Wanda. Going over this stuff in my mind puts me right back in the center of the craziness and sometimes I wonder where it's all going to end up."

"What were you going over in your mind this time?"

"That time when I met Ritter von X was when I had just had an extraordinary metaphysical experience—and one of most frightening psychic warnings one can receive. The warning came true and led to a very sad tragic event. I was actually speaking German with Ritter von X and I don't know a word of German. I saluted him, Wanda."

"It was on a warm summer evening and an old friend who I had known since elementary school, stopped by for a visit. We had not seen each other for years. My daughter, Zephera, had a friend visiting her that evening and my brother, Fred, was also visiting from Montana.

During the evening, I received a phone call shortly before midnight. I recognized the voice immediately but made no outward comment to indicate who had called, or any gestures to attract attention. It was a warning, a deadly psychic warning. In brief, I was told that certain dark forces were about to launch a psychic assault on me that would lead to a tragic death in an auto accident.

The message was less than three minutes. However, long enough to have impacted my life to the point it will remain with me for the rest of my life. I did not express any fear during the conversation and kept the experience quiet after our conversations had ended. Fred immediately picked up on who it was by observing me and we remained silent after the call. I certainly wasn't the right time to discuss the matter.

Within an hour of the phone call, I was saying good-bye to my friend. He lived within thirty miles of my home in the Santa Cruz Mountains. After I closed the front door behind him I explained to Fred the urgent nature of the phone conversation. Because of the source, I gave the warning my fullest attention.

After discussing the conversation with Fred, it was time to give Zephera's friend a ride home. They went to the car and waited for me. As I walked outside the house a mysterious dark object caught my attention just above me in the night sky. I called to Fred. Just above us and below the Redwood treetops was a black circular sphere approaching us. It was about three feet in diameter. It was like a black balloon drifting towards us and as if it knew its direction. You could have imagined what words dropped from our mouths to describe this ominous dark sphere. Once we had focused our full attention on this dark approaching unknown force, it began to slow and stopped within about twenty feet. We continued to use a collection of words (mostly swear words) in describing and discouraging its ominous presence. The sphere began to move back into the night sky towards the north, slightly diminishing until it disappeared from our sight. Meanwhile, Zephera was still waiting in the car. We fastened our seat belts and returned home within thirty minutes, unscathed.

Early the next morning, at approximately 6 a.m., I received a call from the father of my friend who had visited the prior evening. He told me that his son had been in a serious auto accident on his way home. His car had somehow run off the highway and tumbled into a deep canyon. He had been rescued by helicopter and flown to a near-by hospital. He survived the treacherous roll but his arm had to be amputated as a result of the ordeal.

Wanda and I fell silent for a few moments. She then broke the silence.

"And then you actually met him, right?"

"Yes. Remember, I first spoke with him on the phone, after doing some research at the library and finding his name in the Encyclopedia of Associations. I simply called him up and told him about my research on the theories of the Hollow Earth. That started the whole thing with him. He invited me to a meeting. We talked. I got those most recent tapes that you helped transcribe. And now, here we are."

"And now here you are." Wanda is incredulous and sympathetic. "Oh Dear God, when's it all going to end?"

༄ Epilogue ༄

*There are more things in heaven and earth, Horatio,
Than are dreamt of in your philosophy.*

—Hamlet, William Shakespeare

Dear Fellow Truth-Seeker,

It is time for our mission to get underway. The Hollow Earth Research Society (HERS) formally known as the International Society for a Complete Earth (ISCE) is actively planning a reenactment of Admiral Richard Evelyn Byrd's exploration to the North Pole.

To this day, the controversy mounts. Dr. Raimund Goerler, Ohio State University archivist from the Byrd Polar Research Center claims to have found the long-lost diary. Dennis Rawlins, is an astronomer who, after examining it, announced that the famed American explorer Admiral Richard E. Byrd, who claimed to have been the first person to fly over the North Pole in 1926, failed in his attempt, missing his destination by 150 miles.

How ironic were those ten years before the archivists' and the navigational experts' claims of discovery, for it was Ritter von X who told me that "In the near future, even the late great polar explorer, Admiral Richard E. Byrd, will be utterly discredited for his discovery of the North Pole." It was Ritter von X who gave me a copy of the alleged secret missing diary. It was titled *The Flight to the Land beyond the North Pole* or *Is This the Missing Secret Diary of Admiral Richard Evelyn Byrd?*

With his permission, I reprinted the diary in 1990. Two years later, I was contacted by Bolling Byrd-Clark, the daughter of Admiral Byrd. She told me that she remembered her father talking about his diary when she was as young as eleven years old. "It is lost, stolen, or has disappeared," she said. That is how we met.

In the Admiral's book, *Skyward*, written in 1929, just two years after his flight to the North Pole, he also refers to his diary and its contents.

Admiral Richard E. Byrd (1888-1957) revisits his old hut at the site of Little America II. Photo taken in 1947 during the Navy's Operation Highjump expedition.

Few people knew of his diary until Ritter von X gave me a copy for publication.

In April of 1996, at the invitation of Bolling Byrd-Clark, I attended the first Byrd Polar Research Colloquy.

The meeting, held at the Byrd Polar Research Center (BPRC) at Ohio State University, was both a seventieth anniversary celebration of Admiral Byrd's 1926 flight over the North Pole and the fiftieth anniversary of Operation High Jump, the code name for Admiral Byrd's South Pole expedition. Attending the conference was a vast array of scientists, researchers, historians, explorers, and other enthusiasts interested in sharing experiences relating to exploration of the earth's Polar Regions.

Ed Hayes, the university's vice president for research, opened the meeting with a presentation of his views on university participation and the BPRC's impact on future polar research.

Other speakers from OSU, alumni, and invited guests lectured on the latest evidence of global warming and discussed the long-term effects of it on the fragile polar environments.

Colleagues who had traveled with Admiral Byrd on various missions between 1939 and 1955 featured well-illustrated presentations recalling their memorable and daring exploits in some of the harshest regions on earth.

Byrd's Fokker—Admiral Richard E. Byrd's plane on a raft from the "Chantier" to its base, at Spitzbergen, Norway, preparing for his epic flight to the North Pole on the 9th of May, 1926.

New laboratories, archival facilities, and books were all part of the first Colloquy.

Discussions ranged from first-hand experiences with Admiral Byrd to forays into the most remote, inaccessible, and environmentally harsh regions to scientific prognostications by the multinational and multidisciplinary scientific teams at the Byrd Polar Research Center.

Data gathered from recent expeditions, using state-of-the-art research tools, have enabled scientists at BPRC to measure the latest projections and effects of population growth, cumulative environmental pollution, and global cycles of continental plate motion, glaciations, and meteorological change.

Other keynote speakers were Dr. Leonid Polyk, BPRC member and curator of the Marine Sediment Core Repository and Col. Richard Lockhart (retired), who recollected his involvement with Operation High Jump, of which he was a participant. Dr. Richard B. Alley a BPRS alumnus, lectured about the abrupt climate change during the end of the ice age.

A book-signing party was held in honor of Sir Charles F. Passell. His book, Ice, was available for purchase at the conference.

One of the most curious comments came during the archivist Dr. Raimund Goerler's talk on his recent discovery about Admiral Byrd's missing diary. North Pole researcher Charles Passel interjected his thoughts. He said it was a fact also verified by a Swedish expedition that discovered tropical plants and trees floating in fresh water. The expedition also found a slight curvature the farther north they traveled, and even suggested the possibility of a Hollow Earth. Needless to say, laughter rippled through the audience.

Bolling Byrd-Clark interrupted, saying to the crowd, "Well, I think it's best that we just have an open mind here today."

Byrds of a Feather—From left to right: Katherine Byrd-Breyer, Leverett Byrd, Bolling Byrd Clark and Robert Byrd Breyer

I, myself, remained quiet, smiled at Byrd-Clark, and simply listened carefully to the various reactions of the professional scientists and others who knew the Admiral personally and had been with him on his Arctic and Antarctic expeditions.

Saturday's speakers featured Dr. Garry McKenzie and Dr. Ingrid Zabel from the Massachusetts Institute of Technology, who lectured about her radar video (one-year time-lapse photography of the conditions at the South Pole.)

Dr. Zabel's impressive presentation revealed a strange phenomenon that showed what appeared to be a small black hole in the sky that became a large black space and then disappeared. Her only comments concerned the position of the camera during the radar video recording.

Dr. David Bromwich, senior research scientist associated with the

Polar Meteorology Group and member of the BPRC, gave an eye-opening speech and discussion regarding the global impact of polar processes. Captain Brian Shoemaker from the American Polar Society delivered a very informative speech concerning the Antarctic Treaty and the future of Antarctica.

During the final lecture of the conference, Dr. Raimund Goerler, university archivist from the Byrd Polar Research Center, created great controversy and an ensuing debate when he officially announced his startling discovery of "new evidence" regarding Admiral Byrd's North Pole diary.

Goerler explained that Byrd was approximately 150 miles south of that pole when his attempt failed. The information was based on penciled and erased notes that Goerler discovered in January of 1996 in the university's library.

In The Polar Times, published by the American Polar Society in the fall–summer of 1996, archivist, Raimund Goerler, claimed that Admiral Byrd's diary proves that he missed his mark.

Did Admiral Byrd fail to reach the North Pole?

"Famed American explorer Admiral Richard E. Byrd, who claimed to have been the first person to fly over the North Pole, may actually have turned back too soon, missing his destination by about 150 miles."

So begins an article by the Associated Press on the conclusions of Dr. Raimund Goerler, chief archivist of the Byrd Polar Research Center. The article went on:

> The research center, located at Ohio State University, is home to many records, artifacts, and other historical data related to polar research.
>
> Goerler claims that earlier this year he found Byrd's long-lost diary in a mislabeled box of Byrd's memorabilia.
>
> A later interpretation of the diary's navigational notes by Dennis Rawlins, a Baltimore-based navigation expert and polar expedition historian, led Rawlins to believe that Byrd failed to reach the North Pole, even though he claimed to have reached it.

A passage in the Byrd diary allegedly portrayed an erased, but still readable sextant recording that may have put Byrd about 165 miles south of where he claimed to have been later in his official report.

Goerler claimed that neither he nor the Byrd Polar Research Center endorses Rawlins' findings. Goerler said Rawlins' conclusion was based on a sentence written by Byrd stating "we should be at the pole."

To Rawlins, the word "should" implies doubt about an achievement that would, upon return, be asserted as a certainty.

If Rawlins' interpretation is confirmed, then Byrd would lose his standing as "first person to reach the North Pole," to Norwegian explorer Roald Amundsen, who traveled over the pole three days later, on May 12, 1926.

Rawlins did claim that the diary does establish that Admiral Byrd and his pilot Floyd Bennett did indeed leave Spitzbergen, Norway, and attempted [to reach] the pole by flying some 600 miles over the ice in a tri-motor Fokker.

But Rawlins claim is that Byrd's diary is not consistent with the Admiral's making it all the way to the North Pole.

The Polar Times headline reads: "Examination of Admiral Byrd's navigation on his flight to the North Pole," by William E. Molett, Lt. Col. USAF (retired).

A long-lost diary of Admiral Byrd was found recently by Dr. Raimund Goerler, archivist at the Byrd Polar Research Center in Columbus, Ohio.

On page eleven of this diary is an erased figure which can just be made out to be what a sextant altitude is, apparently.

Based on an analysis of this single erased figure, Dennis Rawlins has announced to the world that Byrd made a fraudulent claim to have reached the North Pole.

According to Rawlins, Byrd was 146 nautical miles short of where he thought he was at the time of this supposed sextant observation and, continuing on, he turned around over 100 miles short of the pole. This author [Lt. Col. William Molett] maintains the erased sextant reading was just that—a figure that Byrd did not use because he knew it was incorrect.

Richard Byrd took and recorded a sextant reading of 18 degrees, 15 minutes, and 30 seconds at Greenwich Civil Time of 7 hours, 7 minutes. For the same time at another place in his records, he recorded an altitude of 19 degrees, 22 minutes, and 14 seconds.

The 19+ degree altitude has been erased but still can be read. If this erased figure was a true altitude at the recorded time, then Byrd manufactured ground speeds up to this time and all ground speeds and sextant readings after this time. If this erased figure is ignored, then everything he wrote about the trip appears to be genuine.

In using a bubble horizon sextant, it is essential that the plane is flying straight and level at the time of the observation. Even small turns with the aileron or rudder can make for large errors in the sextant reading.

For several years, Byrd had used this sea sextant, which he had modified for aerial use. In the air, because of aircraft motion, a single reading of the sextant cannot be relied on for accuracy.

In the case of the 7:07 reading, the pilot (Floyd Bennett) may have made a small turn to correct to course, perhaps without even realizing he had done so, and caused Byrd to make an erroneous reading.

When Byrd wrote down the 19+ altitude, he realized immediately from his position that it was wrong, and probably went back to an earlier reading of the set of readings [and] erased this error.

It is also possible that he first misread the altitude and then with another look at the sextant, corrected the reading; this is a normal procedure to check one's readings.

Whatever the reason, this erased altitude was not used by Byrd. All of the other readings were taken with apparently excellent accuracy. Byrd reported the air was not bumpy. Air over the Arctic Ocean is possible the smoothest in the world.

There are no mountains to give up and down movement the air, and with the sun shining 24 hours a day, night temperatures and day temperatures remain much the same, which also contributes to smoothness in the air.

Byrd's sextant readings seem almost too good to be true, but smooth air and careful observations of the sun would give him pretty accurate readings.

The erased sextant reading in Byrd's diary played no part in Byrd's navigation.

This author is awed by Byrd's careful planning and execution of his first flight to the North Pole. He went north on the 11E meridian. He planned his takeoff time to give him his course by the sun lines in the mid-portion of the flight and latitude by sun lines as he approached the Pole.

About one hour after he started south, the sun crossed the 15E meridian directly at the sun. The sun's shadow was directly down the middle of his sun compass, indication he was exactly on his desired course. That he was and remained on his course was confirmed when, at about 120 miles north of Spitzbergen, he spotted Grey's Point almost dead ahead.

Byrd helped design the drift meter he used for drift and ground speed, [and] it is inconceivable that he could travel for nearly six hours and not detect that his actual ground speed was only 53 knots when he consistently computed it to be around 77 knots or higher.

Byrd made huge contributions to naval aviation and aviation in general. He was probably one of the ten best navigators in history.

When he said he reached the North Pole in his tri-motor aircraft, you can believe it despite the criticism by non-navigators.

Those who knew Admirable Byrd

To the editor of the Polar Times:

The last time Dennis Rawlins purported to produce irrefutable evidence that an explorer had lied about his feat, the evidence turned out to be only the serial numbers on Admiral Peary's chronometer. Rawlins goes to great lengths to debunk the extraordinary accomplishments of famous polar explorers.

—Gilbert M. Grosvenor
Chairman of the Board
National Geographic Society

The accomplishments of great men are often belittled by would-be hero's who present the negative side of life instead of the positive. The media finds it so easy to criticize and difficult to compliment.

Admiral Byrd's story and reputation of being a good leader, a good navigator, and a good explorer has been questioned by individuals who strike below the belt. Will the professional disparagers of the future try to negate the flight to the moon as they have targeted the life of Admiral Byrd, especially the flight to the North Pole?

For four years I worked for Admiral Byrd—to train his sledding teams, to accompany him on his first Antarctic expedition (1929-30), to prepare his supplies for the second expedition (1933–34), and to accompany him on many of his fund-raising lectures. At no time did I find him a liar or even drunk as reported in the New York Times on 12 May 1996.

On the contrary, he was always a gentleman, lived up to the good name of the United States Navy, was proud of his family, and more than anything else was an inspiration and strength behind all Antarctic exploration, discovery, and scientific development. By his leadership and selection of good men like Dr. Lawrence M. Gould of the University of Michigan senior scientist and second in command of the first Byrd Antarctic Expedition recorded the geographical results of the Byrd Antarctic Expedition. Admiral Richard E. Byrd's place in the world's history of exploration is at the top.

—Norman D. Vaughan

I have read the article in the New York Times disparaging the character of Admiral Byrd. Having been a member of the Byrd Antarctic Expedition II, 1933–35, as a seaman on the BEAR, I say it is a lot of baloney.

Fear of flying!! Many people today would fear flying in the Arctic and Antarctic with equipment likely to fail, no Navy help, no reliable weather route and no backup aircraft to search for him if he was forced down. He made the flights in spite of these possibilities. If he was afraid of flying, he never showed it.

Terrible navigator!! He was the foremost navigator of his day, a man who did a lot to advance aviation, as well as explore the Polar Regions. With his primitive equipment he found his way out and back to safely. What is so terrible about that?

Did not pilot his plane!! Amundsen, Ellsworth and Wilkins did not pilot their planes. All of them picked the best pilots they could find.

What's wrong with that? Byrd bashers should try to duplicate his accomplishments before they try to criticize him.

—Gordon Fountain

My personal memories from the Byrd Polar Research Conference

After the conference, Bolling Byrd-Clark, Admiral Richard E. Byrd's daughter, invited me to dinner with ten other Byrd Polar Research members. The get-together included Bolling's sister, Catharine Byrd Breyer, and a nephew, Leverett Byrd.

I was fortunate to be seated between Byrd-Clark and her nephew, who, incidentally, looked identical to a younger Admiral Byrd.

As we dined, Byrd-Clark offered a toast, saying, "Here's to the old man, himself" in what indeed was a tremendous reunion of family and history as well.

During the affair, I had the honor of being introduced to the group by Byrd-Clark as a long-time acquaintance of hers, as well as a Byrd researcher whose interests also included researching the theory of a Hollow Earth.

Byrd-Clark continued, emphasizing her belief that an open mind is essential regarding the Hollow Earth theory, and I then had the wonderful opportunity of formally addressing many questions posed by the other guests. I shared Icelandic and Tibetan legends and spoke of my many journeys to Iceland, where, for a time, I had made my home.

After discussions that varied from my research of Iceland to the many contacts and friends I had made during my time there, I moved on to introduce the legends of Odin and Thor. Odin represented the All-Father God who dwells in the center of the earth. This place is the Icelandic legendary world called Asgarth, where Odin's son Thor also lives in the Citadel of the Gods.

The discussions carried into the late hours of the morning, by which time no one in attendance denied the possibility of a Hollow Earth. Their interest was piqued at the mysterious conversations I had with Ritter von X about the late Admiral Byrd.

To continue, dear researcher, you may well imagine that this journey will require some time yet to arrange. However, we are actively getting the process underway. This is an enormous undertaking with many logistical considerations. This is the main objective and purpose of HERS, with the ultimate goal of proving or disproving the theory of the Hollow Earth.

Many people throughout the world have been researching this subject matter for years. In addition to many accredited researchers and investigators, other individuals have compiled masses of information, literature, tapes, and so on. For an example, take yourself. How did you first hear about the theory of the Hollow Earth? What intrigues you about it? Do you want more information? Did you receive this information from a friend, or, perhaps by a metaphysical contact? Do you have sufficient or conclusive evidence proving this theory? Do you believe the Hollow Earth exists? Does merely believing support the facts? What, in your estimation, constitutes fact? Can anyone say with irrefutable proof that the Earth is hollow?

I am sure you will agree that theories certainly suggest the Earth is hollow. The possibility, coupled with so much corroborating data, is strong enough to support the inquiries on the subject. But as a researcher or investigator, are you completely satisfied with all the information and materials that have been brought to your attention by your own efforts? Participation in this organization is an attempt to go beyond the theory itself. A theory does not determine fact but produces an educated guess about the possibility—all brought about by researchers from all walks of life and professions. The facts remain to be verified.

If you knew for certain that the Earth was hollow and inhabited by a superior strain of beings that are thousands of years advanced in technology, science, and mathematics, in an environment that supports life indefinitely, would that change your life and the world in which you live?

Nearly everyone who has heard of this theory has read Dr. Raymond Bernard's book, *The Hollow Earth*. He has dedicated his book to the future explorers, who he hopes will actually discover it. It is the Hollow Earth Research Society's intention to become these new world explorers.

This is a personal invitation. It is for those individuals who want first-hand information and experience of an actual flight to the North Pole.

If you contact the author at *www.hollowearthresearch.com*, the organization will keep you informed of the progress and itinerary along with providing additional articles of interest, new information and research, photos, opinions, and resources. The society is the clearing house for information dedicated to seeking the facts. All research is posted on the website at *www.hollowearthresearch.org*. This is the vehicle to disseminate that information and plan exploration. One such venture is the Byrd trip reenactment. Our goal is to investigate the possibility of a Hollow Earth and its surrounding mysteries and to verify whether Admiral Byrd in fact reached the North Pole.

Captain John Cleves Symmes's Globe—John Cleves Symmes's handmade wooden globe shows the earth's interiorand accessible through the vast opening at the poles

The first stop is planned for Washington, DC, at the Smithsonian Institute. Did you know that the Smithsonian Institute was founded because of the United States expedition of 1838–1842 led by Lt. Charles Wilkes?

A short film in the Natural History Museum gives visitors a history of the expedition from its origins early in the nineteenth century. It embraces John Cleves Symmes's belief in a Hollow Earth. His wooden globe illustrates his "Holes in the Poles" theory. His ideas were not widely accepted, but his call for an exploratory expedition was supported by New England whaling and commercial interests, patriotic citizens, and naval officers. Congress authorized a Pacific exploration in 1828. After a decade of controversy and planning, the ship sailed from Hampton Roads, Virginia, on August 18, 1838.

We invite you to the first-ever Hollow Earth Convention!

Do you have something to share?

HERS and the Conventions Committee in Hamilton, Ohio, are planning the first world wide "Hollow Earth Convention" in honor of Captain John Cleves Symmes.

The director of HERS encourages all those who may wish to participate and/or contribute to the effort to contact him. If you have any special research, workshops, presentations, stories, artifacts or anything you wish to share or would like to be a part of the convention, either as a contributor or a visitor, please contact HERS. HERS will respond to all those who contact the director with continuous updates regarding the progress of this proposed convention. There are literally thousands of dedicated researchers, UFO organizations, scientists, pilots, biblical scholars, authors, university professors, and many psychics, metaphysical people and others of all walks of life around the world have something to share.

Why Hamilton, Ohio?

In 1826, Hamilton, Ohio resident Captain John Cleves Symmes developed the hollow earth theory. In fact, he further regarded the theory as the "Concentric Ring Theory." He was an ex-army officer and a business man. He believed the theory in nearly the same way Dr. Raymond Bernard except for his "Concentric Ring" explanation.

John Cleves Symmes

Symmes dedicated most of his life advancing his theory and raising money to support an expedition to the North Pole to find the entrance to the hollow earth. In 1818, with a number of growing enthusiasts, Symmes shared his ideas while on an international lecture tour. In a letter addressed to "All the world," and directed to various publications, politicians, learned societies, and heads of state throughout America and Europe, he wrote:

> LIGHT GIVES LIGHT, TO LIGHT DISCOVER—"AD INFINITUM."
>
> ST. LOUIS, (Missouri Territory,)
> *North America, April 10, A. D. 1818.*
>
> TO ALL THE WORLD!
>
> I declare the earth is hollow, and habitable within; containing a number of solid concentrick spheres, one within the other, and that it is open at the poles 12 or 16 degrees; I pledge my life in support of this truth, and am ready to explore the hollow, if the world will support and aid me in the undertaking.
>
> *Jno. Cleves Symmes*
>
> *Of Ohio, late Captain of Infantry.*
>
> N. B.—I have ready for the press, a Treatise on the principles of matter, wherein I show proofs of the above positions, account for various phenomena, and disclose Doctor Darwin's Golden Secret.
>
> My terms, are the patronage of this and the new worlds.
> I dedicate to my Wife and her ten Children.
> I select Doctor S. L. Mitchell, Sir H. Davy and Baron Alex. de Humboldt, as my protectors.
>
> I ask one hundred brave companions, well equipped, to start from Siberia in the fall season, with Reindeer and slays, on the ice of the frozen sea; I engage we find warm and rich land, stocked with thrifty vegetables and animals if not men, on reaching one degree northward of latitude 82; we will return in the succeeding spring. J. C. S.

Symmes's Letter to Congress

Following is an article from the Hamilton's Journal News, written on March 14th, 1991. It describes a gathering of a group of Hamilton civic and city officials who sheltered themselves from the icy rain with umbrellas as Sonia August, chairman of the restoration project for the John Cleves Symmes Hollow Earth monument delivered a few remarks at the memorial's re-dedication.

Hollow Earth Solidarity

"About 80 people greeted the official return of the Symmes Monument as Historic Hamilton Inc. re-dedicated the 150 year-old grave marker. The monument has been under reconstruction for about 8 months and was returned Monday from Hamilton sculptor, Edgar Tafur. Tafur used copper tubing to reinforce the sphere on top of the monument and the corners and resurfaced the foundation.

Workers hoisted the monument back onto its resting place earlier this week. Representatives for Historic Hamilton Inc. and from the neighborhood spoke following the dedication by Historic Hamilton's president, Nancy Tryloff. Susan Vaalere, co-chairman of the restoration project, gave additional history about the graveyard and Symmes' theories.

Hamilton mayor, Adolf Olivas also spoke, saying that although Symmes' theories seem farfetched to the present generation, ideas just as farfetched circulate today. "Having the monument here brings further shine to Hamilton's star," he said, "and we'll do our best to make sure, for the next 100 years, that it is preserved."

Captain John Cleve Symmes' Monument, Hamilton, Ohio—Photo taken by Audrey Morrison of Lane Public Library Reference Services Team

It was only after his death that one of his ardent followers, a newspaper editor named Jeremiah Reynolds, helped influence the US government to send an expedition to Antarctica. The idea gained support by President John Quincy Adams and, in 1828, Congress approved the expedition. The ship sailed in 1838. The artifacts from this expedition became the foundation of the Smithsonian Museum in Washington, DC. Is it ironic that our national museum was founded as a result of a congressional order to search for the entrance to the "hollow earth?"

As mentioned, HER main objective is to prove or disprove the Hollow Earth theory. If you wish to be involved in this project by participating in the proposed journey and/or the ongoing research, you are invited to combine your interests with HER's. We encourage you to visit our website, and to share your questions, comments, and assertions.

The second stop we have planned will be Reykjavik, Iceland, which, interestingly, was the setting for Jules Verne's novel, *Journey to the Center of the Earth*. We will also visit Mt. Snaefellsjokull, where the famous movie was filmed. We will visit museums and explore the Nordic legends of Odin, Thor, the Valkyries, and Asgarth, the Icelandic legendary city in the center of the earth.

Reykjavik, Iceland

Further, we will explore how World War II German leaders were inspired and perhaps driven by these legends, and we will examine how German composer, Richard Wagner's research on this topic profoundly influenced his epic four-opera cycle, *The Ring of the Nibelung*.

Third, the actual flight to the North Pole will originate from Spitsbergen, Norway, as did Admiral Byrd's. We cannot predict what will be seen and experienced. It will be an attempt to gain independent verification of Admiral Byrd's lost diary contents, and give those interested a first-hand experience to make their own judgments on this subject.

Our base camp at Spitzbergen, Norway

This reenactment journey will be an attempt to verify the recent claim that Admiral Byrd missed the North Pole altogether and to clarify the mysteries that have shrouded him for over fifty years.

Care to have a great adventure that may be "out-of-this-world" in the real sense? We invite you to join our research team—we hope to hear from you soon.

This is where the story ends and the journey begins.

Illustration Credits

Opposite Dedication— Gerhardus Mercator's North Pole map
Image source: http://www.helmink.com

Opposite p.1— Greenland Map (with and without ice)
Image source: Published by the National Geographic Magazine Washington DC
Quotation: Mercury: UFO Messenger of the Gods, by W.D. Clendenon, published 1991, page 108

p.9— Edmond Halley
Image source and quotation: http://www.dioi.org/kn/hollow.htm

p.23— Photo of Vietnamese Buddha
Image source: Given to me by a friend in South Viet Nam in 1968 during my tour there.

p.47—Calvary Dusk. Image source: www.dreamstime.com

p.63— Photo of Ritter von X
Image source: Given to me personally during my first meeting with him.

p.71— Letter in German
Image source: Reproduction of letter mailed to Ritter von X in 1947 from K.C. (confidential) about U-Boat 209. Personally given to me by Ritter von X.

p.182— Cutaway of the Hollow Earth
Image source: Michael Abbey of Visual Communications P.O. Box 357, Santa Cruz, CA 95061

p.188— North Pole
Image Source: http://earth.jsc.nasa.gov/ then go to:
ftp://eol.jsc.nasa.gov/ISD_highres_AS16_AS16-118-18885.JPG

p.189— Northern Lights Ring
Published in the German magazine *Zeiten Schrift*. Issue 1 November 1993. Photo taken by Dave Fritts.

p.189— Ozone Ring, from *Popular Science* January 1992. Taken by the Nimbus-7 satellite.

p.210— St. Germaine
Image source: (Public Domain Image)
Quotation: The Brotherhood of the Rosy Cross by Arthur Edward Waite (Printed by William Rider & Son Limited, London, 1924)
Saint-Germain and Cagliostro
http://www.hermetics.org/brc-17.html

C.A.Vulpius. *Curiostäten der Literarisch Historischen Vor und Mitwelt*, 1818, pp. 285, 286

p.214— Mt Shasta
Image source: Photo taken by my daughter Zephera.

p.265— Atlantis Illustration: by Lloyd K. Townsend
Image source:
http://en.wikipedia.org/wiki/File:Townsend_Lloyd_K_-_Atlantis.jpg
Quotation: http://www.gutenberg.org/ebooks/1571

p.284— Sveinbjorn Beinteinsson
Image source: Photograph by Jónína K. Berg
http://en.wikipedia.org/wiki/File:Sveinbj%C3%B6rn_Beinteinsson_1991.jpg

p.296— South Pole Oasis
Image source: Published by the National Geographic Magazine October 1947 VOL 92 No. 4 Washington, DC

p.297—South Pole Lake – Plane
Image source: Published by the National Geographic Magazine October 1947 VOL 92 No. 4 Washington DC.

p.301—Admiral Byrd
Source: http://www.veterantributes.org/TributeDetail.asp?ID=398

p.303— Byrd's Fokker
Image source: Published by the National Geographic Magazine October 1947 VOL 92 NO 4 Washington, DC

p.304— Katherine, Levrett, Bolling and Robert
Image source: Photo taken by Danny Weiss while attending Admiral Byrd's 70th anniversary at the Byrd Polar Research Society on April of 1995.
Tribute to Bolling Byrd-Clark
http://www.cookpolar.org/bollingbyrdclarke.htm

p.312— Symmes Globe
Image source: Magnificent Voyagers Herman J. Viola and Carolyn Margolis, Editors Smithsonian Institution Press, Washington, D.C., 1985 page 11

p.313— Symmes Portrait
Image source: Magnificent Voyagers Herman J. Viola and Carolyn Margolis, Editors Smithsonian Institution Press, Washington, D.C., 1985 page 11

p.314— Symmes Congressional Letter

Image source: Magnificent Voyagers Herman J. Viola and Carolyn Margolis, Editors Smithsonian Institution Press, Washington, D.C., 1985 page 10

p.315— Symmes monument rededication photo and article
Image source: Photo taken by Audrey Morrison of Lane Public Library Reference Services Team, Hamilton, Ohio 2009
Article by Journal News of Hamilton, Ohio 14 March 1991

p.316— Reykjavik, Iceland
Image source: Photo courtesy of Patrice Raplee
http://www.offbeattravel.com/iceland-reykjavik.html

p.317— Spitzbergen – Longyearbyen
Image source: http://www.pri.org/theworld/?q=node/16567&answer=true